BOTTLED LIGHTNING

BOTTLED LIGHTNING

SUPERBATTERIES, ELECTRIC CARS, AND THE NEW LITHIUM ECONOMY

SETH FLETCHER

HILL AND WANG

A DIVISION OF FARRAR, STRAUS AND GIROUX   NEW YORK

HILL AND WANG

A division of Farrar, Straus and Giroux

18 West 18th Street, New York 10011

Published in 2011 by Hill and Wang

First paperback edition, 2012

The Library of Congress has cataloged the hardcover edition as follows:

Fletcher, Seth, 1979–

Bottled lightning : superbatteries, electric cars, and the new lithium economy / Seth Fletcher.

p. cm.

Includes bibliographical references and index.

ISBN 978-0-8090-3053-8 (hardback)

1. Lithium cells. 2. Lithium industry. 3. Electronic industries. I. Title.

TK2945.L58 F58 2011

621.31'2423—dc22

2010047695

Paperback ISBN: 978-0-8090-3064-4

Designed by Abby Kagan

www.fsgbooks.com

P1

For Leigh

It is quite possible that the man who has taught us to put up electricity in bottles has accomplished greater things than any inventor who has yet appeared.
— *The New York Times*, June 11, 1881

CONTENTS

BOTTLED LIGHTNING

PROLOGUE

To skeptics, the Chevrolet Volt concept car was nothing but a public-relations stunt, an unserious stab at building green credibility through future-shock styling and a hollow promise of untold advances in fuel economy. After all, the annual North American International Auto Show more closely resembles a Paris fashion show than tent day on the dealer lot. Every January, Detroit's Cobo Hall is filled with outlandish concept cars—hydrogen-fuel-cell cars whose time is always just another year away, hallucinatory "design studies" whose lines might one day help determine the grille pattern on an otherwise bland sedan. Nothing wrong with that: it's the way the automotive industry works. The artists and research-and-design people have their fun, and then the executives and engineers water the fanciful down to something safe, something that the board suspects might actually sell. Sometimes the technological advances that run down the automotive catwalk do eventually appear in the real world. But in early 2007 GM had a credibility problem. It was natural to wonder how serious the manufacturer of the Hummer could be when it said it would soon build an entirely new type of twenty-first-century automobile.

The skepticism surrounding the Volt was partly informed by the documentary *Who Killed the Electric Car?*, which had premiered at the Sundance film festival the previous year and had quickly become a cult hit, delivering to an unexpectedly large audience footage of General Motors' seizing all but a few specimens of its earlier EV1 electric car from their heartbroken drivers, hauling them out to the desert, and crushing them en masse behind a security fence. The documentary didn't directly accuse GM of conspiracy, but it didn't matter. Particularly among the young and the environmentally conscious, a new piece of conventional wisdom threatened to crystallize: General Motors, an evil American corporation, had colluded with the oil companies (also evil) to rid the land of an existential threat, the electric car.

Yet the truth was that by the day of the auto show, General Motors had done vastly more production-intent work on the Volt than they ever would for a normal concept car. It hadn't been easy, but its product-planning chief, Bob Lutz, had secured a good-faith commitment from his superiors to build the Volt. It was to be a moon shot, a game changer— an entirely new type of car, one more relevant and useful than any electric vehicle that had come before. Thanks to its series hybrid drivetrain—a battery-powered electric traction motor backed up by a gas-powered generator—the Volt would hardly ever use gasoline around town. If you weren't opposed to buying gas, however, you could also drive the car cross-country. And the Volt's power architecture could be transferred to any number of other vehicles.

The Volt, or at least the idea of the Volt, was promising enough that some of GM's critics—even some of the people behind *Who Killed the Electric Car?*—allowed themselves to be cautiously open-minded. There were indications that this time, General Motors might be serious. Rumors had been spreading for months that the company might be ready to get back into electrification. Lutz had been stung by *Who Killed the Electric Car?* More important, he had come to see GM's ceding of the lead in the newly fashionable hybrid market to Toyota and its Prius as a terrible mistake. The righteous glow of the Prius was by 2006 shining on every car and truck Toyota built. The Prius was the ultimate image booster for the company that appeared set to surpass GM as the world's largest automaker. Then, the arrival of start-up Tesla Motors and its glamorous electric roadster finally made Lutz snap.

The skeptics were right about one thing: the Volt was definitely about image. The car was Lutz's baby, designed in pseudo–Skunk Works fashion, pitched as a reputation rebuilder for a stumbling, increasingly unprofitable company. He had given the designers little direction except that the car should be a technology showcase, and that it should be an electric—not a hydrogen-fuel-cell car, which by 2006 GM had spent at least a billion dollars researching. The prototype that the designers came up with was sexy, like a twenty-second-century Camaro, and on the day of its unveiling, standing beneath the blue-gelled stage lights next to his gleaming car of the future, Lutz appeared to gloat.

Announcing the Volt, General Motors executives argued that the time for the electrification of the automobile had finally arrived, thanks to a critical enabling technology: the lithium-ion battery. This was the same technology responsible for the miraculous shrinkage of the cellular phone. These batteries were three times more energetic and less than half the weight of the lead-acid cells that drove the EV1. Make that the lead-acid cells that *doomed* the EV1. The company line was that heavy, inefficient lead-acid batteries, the best available in 1996 when production of the EV1 began, were so bulky that it was impossible to shove a backseat into the car, so short-lived that the vehicle was limited to a maximum range of one hundred miles or so, after which, if you ran out of juice, you were, as Lutz would later say, "truly screwed." This is why the EV1 failed, GM contended: it was really just an inadequate car. But the lithium-ion battery changed everything. It could bottle up enough energy in such a small and light package that suddenly an affordable four-passenger electrified vehicle was feasible.

There was a major caveat: the batteries weren't quite ready. The type of lithium-ion battery used in laptops and cellular phones wasn't suitable for use in a car, experts argued. It wasn't safe enough, powerful enough, or durable enough for the many years of abuse that an electric-car battery would have to endure. That could change quickly, however: around the world major electronics companies, venture-backed start-ups, and scientists in national and university labs were developing new strains of lithium-based batteries that had the potential to make the Volt reality. Once the batteries were ready, the car would be ready too.

Plenty of people thought GM was simply setting itself up, once again, to blame the failure of an electric-car program on the batteries. Even the optimists were afraid of being fooled again. This new battery technology was promising, and the Volt's blended power train cleverly circumvented the most common obstacle to electric drive—limited driving range. None of that would necessarily stop GM from milking the car for PR value without ever putting it into meaningful production. But there was a chance. If the events of the next few years unfolded in just the right way, the electric car's time could finally arrive.

Three years after the Volt's unveiling, I found myself standing on a rocky bluff in a remote and desolate corner of the Bolivian altiplano. In the distance, like an endless vanilla glaze on a mud-chocolate landscape, was the Salar de Uyuni, the largest salt flat in the world, which, by some estimates, holds half of the world's easily accessible lithium. Below lay the construction site of the Bolivian government's pilot plant for lithium production, a small collection of shacks and half-built brick-and-timber structures that had become a source of international fascination and, in Bolivia, plenty of domestic consternation. Delegations from the South Korean government, Japanese and French corporations, and others had been here before me, eager to secure access to Bolivia's lithium riches in preparation for a postoil future. None of them had much luck. Bolivia's president, Evo Morales—Aymara Indian, head of the Movement Toward Socialism Party, former leader of a coca growers' union—had declared that in order to prevent foreigners from stealing the country's lithium, the government would go it alone. This pilot plant was Bolivia's first step toward building a national lithium industry.

Fears about the price and future availability of oil, the increasingly undeniable reality of anthropogenic climate change and the concomitant likelihood of restrictions on carbon emissions, the rise of resource-hungry India and China, the collapse and rebirth of Detroit's auto industry, a dramatic realignment in American politics, the greatest financial catastrophe since the Great Depression and the remolding of the American economy that the crisis seemed to make possible—this strange constellation of circumstances had brought me to this remote corner of the world. Some combination of these factors had by 2010 led nearly every major automaker to announce some level of commitment to the electrification of the automobile. Those commitments caused a rush to supply the batteries the cars of the future would need. That rush initiated a global race to secure access to the world's supply of lithium— the element that is the yeast in the dough of the world's most advanced batteries. All this happened with astonishing quickness. It was fascinating to watch, and for those involved it was by turns thrilling and brutal, an opportunity for a few winners and survivors to change the world and maybe make a billion dollars in the process. Yet it only makes so much sense to speak of these events in the past tense. The real upheaval is only beginning.

THE ELECTRICIANS

Before the invention of the battery in the first year of the nineteenth century, electricity as we know it today—as a stream of electrons that can be made to do our bidding—didn't exist. Electricity was part parlor trick, part mystery. It was a fuzzy force field that could be conjured by rubbing a plate of glass with fur. It was in no way useful, and it wasn't even remotely understood. Only after the battery gave mankind a reliable source of electricity did that really begin to change.

We've known about what is now called static electricity since around 600 B.C., when the Greek philosopher Thales of Miletus began puzzling over a strange property of amber: when rubbed with cloth, amber (called *elektron* in Greek) would, through some invisible mechanism, pull feathers toward itself. The phenomenon resembled magnetism, which the Greeks had observed in iron-bearing stones found near the city of Magnesia on the Meander. After Thales, however, more than two millennia would pass before human understanding of these two forces advanced appreciably. At the beginning of the seventeenth century, some momentum returned when William Gilbert, chief doctor to Queen Elizabeth I of England, discovered that a variety of materials could be electrified by friction, just like amber. Gilbert is the one who coined the word "electricity," drawing on the Greek word for amber to give a name to the force that he called "electrical effluvia." After that, in part because there was no good way to store electricity for use in experiments, the "electrickal arts" progressed only haltingly for the next century and a half.

Then came the Leyden jar. Invented in the town of Leyden, Belgium, in the 1740s, it was literally a jar for electricity. Its walls coated inside and out with metal, the jar was filled with water and then, via a metal chain that dangled down through the lid, charged by an electrostatic machine. (We now know that the Leyden jar is a capacitor, a device that stores charge between a pair of conductors.) One Leyden jar, or several of them wired in series, could hold a significant charge, as Benjamin Franklin learned around Christmas in 1750, when he accidentally hit himself with a charge he'd been building up to kill a holiday turkey. He called it a "universal blow" through the body, which left "a numbness in my arms and the back of my neck which continued til the next morning but wore off." Yet the Leyden jar's usefulness was limited by the fact that it could dump its charge only in that kind of instantaneous jolt. This restricted the kinds of experiments scientists—or, as they often referred to themselves, electricians—could perform, and by the second half of the eighteenth century the true nature of electricity was still a mystery. In 1752, when Franklin performed his legendary kite experiment and determined that the electricity generated by friction was the same stuff as lightning, it was an important breakthrough. But what was *that* common force? No one knew.

The battery was the accidental fruit of a dispute between two Italian scientists over this question. In one corner was Luigi Galvani, a physician at the University of Bologna, who noticed that under certain circumstances, touching a scalpel to the crural nerves in the thigh of a dissected frog caused the legs to kick to life. Galvani came to believe that within the muscles of all living creatures flows an electrical fluid, an "indwelling electricity" generated by the brain and pumped through the body as a motivating force.

In the other corner was Alessandro Giuseppe Antonio Anastasio Volta, a professor of physics at the University of Pavia. Volta had long been interested in the general project of eliminating superstition through the careful study of phenomena still commonly attributed to magic. He thought deeply about the concepts of mind and the soul, and for a while he entertained Galvani's theory as a possible explanation for the relationship between the "will" and the motion of the body. But that didn't last long. Through his research with electrical instruments he became convinced that there was no such thing as animal electricity. Instead, electricity was set in motion by the contact of different metals. When a disembodied frog leg kicks in the presence of electricity, that's because it's a good conductor, just like the human tongue, one of Volta's favorite experimental tools.

Galvani and Volta sparred over the nature of electricity for years beginning in 1792, trading jabs in letters and books. A decisive round began in 1797, when Galvani published a long book devoted to destroying Volta's theory of metallic electricity. Volta could easily handle all of Galvani's arguments but one, which involved a freak of nature that seemed to validate everything Galvani believed in: the torpedo fish, a bottom-dwelling ray conveniently equipped with an organ capable of creating electrical shocks strong enough to kill a man. Galvani believed that some kind of electrical fluid was cooked up in the fish's brain and then piped throughout its nervous system, and he intended to prove it experimentally.

Volta knew that the torpedo fish threat had to be dispatched quickly. He learned how to do so when he read a paper by the English chemist William Nicholson, which proposed that the torpedo fish produced electricity not through its brain, nerves, or will, but through an organ that could be modeled mechanically. Volta ran with Nicholson's idea, determined to build a device that would draw electricity only from the contact of different metals. After only a few months he emerged from his lab with a column of little sandwich cookies, each one a zinc and copper disc separated by brine-soaked cardboard. On March 20, 1800, Volta wrote to Sir Joseph Banks of the Royal Society in England, announcing his discovery of "electricity excited by the mere mutual contact of different kinds of metal." The battery had arrived.

News of the battery spread across Europe as quickly as the infrastructure of the day would allow. Letters describing the new device sailed to England, France, Denmark. Electricians throughout Europe began replicating Volta's experiment, and soon they began building larger and more powerful batteries. Nicholson built one and used it to create what the historian Giuliano Pancaldi described as "loud detonations, clouds of bubbles, gleams of light, shocks felt by up to nine people holding each other by the hand, and a ramified metallic vegetation, nine or ten times the bulk of the wire, when the wire was kept in the circuit of the battery for four hours." Almost immediately the battery enabled major fundamental scientific discoveries. Within weeks, Nicholson and his colleague Anthony Carlisle had used the battery to break water down into hydrogen and oxygen, proving that water was not, in fact, an irreducible element.

Volta called his invention the "organe electrique artificial." Nicholson called the device the "pile," referring to the fact that it is simply a pile of metal and cardboard. Soon, however, the word "battery" emerged in common usage, a reference to the practice of connecting a "battery" of Leyden jars in series to supply electricity.

The battery assured Volta a place in the pantheon. It was "the last great discovery made with the instruments, concepts, and methods of the eighteenth-century electricians," a device that "opened up a limitless field" that "transformed our civilization," wrote the historian John L. Heilbron. The nineteenth-century physicist Michael Faraday, often considered the most brilliant experimentalist in history, called the battery a "magnificent instrument of philosophic research." Auguste Comte, the founder of positivist philosophy, called Volta "immortal" and put him on the *Positivist Calendar*, a proposed reform calendar that celebrated history's greatest thinkers. According to the historian of science George Sarton, the battery "opened to man a new and incomparable source of energy."

Volta earned such effusive praise because of the battery's enduring, history-bending influence. Throughout the nineteenth century, the battery powered the experiments that finally allowed human beings to put to work the amber-borne force field that had mystified thinkers for millennia. The famed English chemist Humphry Davy used large batteries to break various minerals into previously unknown elements—potassium, sodium, magnesium, calcium, barium, strontium. In Copenhagen in 1820, Hans Christian Oersted noticed while giving a lecture that current flowing from a battery changed the direction of a compass that was sitting nearby. Soon, Oersted proved that electricity could induce magnetism. Oersted's discovery led to James Clerk Maxwell's equations describing the relationship between electricity and magnetism—electromagnetism—which led to the electric motor, the generator, the telephone, and every other electrically powered device ever invented.

By the middle of the nineteenth century, the battery found use outside the lab, primarily as a power source for the telegraph. As the battery steadily improved, its uses grew. In 1859, the French physicist Gaston Planté achieved a major breakthrough: the first practical rechargeable battery, a primitive version of the lead-acid cells we still use to start our gas-powered cars. In 1881, the French chemical engineer Camille Alphonse Faure came up with a practical method for manufacturing lead-acid batteries. Soon a shady bunch of European patent scavengers and stock manipulators were trying to get rich on Faure's invention, inflating the small-scale equivalent of a nineteenth-century dot-com bubble, and temporarily giving the battery business a bad reputation. But that didn't stop the spread of the new technology. By the beginning of the twentieth century, lead-acid batteries were widely used to power telegraphs, manage the electrical load in electrical-lighting substations, and support electrical streetcar networks. By then, many of them were also driving cars.

At the beginning of the automobile age, cars powered by gasoline, electricity, and steam all shared the road, and none was an obvious winner. Actually, electric cars had a strong early advantage. They were clean, quiet, and civilized. Gas-powered cars were unreliable, complicated, loud, and dirty. They could be started only with a firm turn of the starting crank, and when that crank backfired it was extremely effective at breaking arms. When they weren't breaking down or inflicting pain, however, gas-powered cars offered something that electric cars couldn't—decent driving range, extendable within minutes with a tin of gasoline from the general store.

Thomas Edison loved the idea of the electric car. Electric cars were a natural, stabilizing, money-generating appendage to the electrical network he had spent his career building. Widespread adoption of the electric car would help sustain his direct current (DC) standard, because charging a battery from an alternating-current (AC) network required an additional piece of equipment, an AC-DC converter. He knew that battery technology would determine whether electric cars would thrive or lose out to the rapidly improving gas-powered car, and he happened to be looking for a new conquest. He had already made, lost, and remade a fortune—already invented the stock ticker, the lightbulb, the phonograph, and the motion picture. He had just closed down a disastrous attempt at mining iron ore in western New Jersey. And so in 1898, he began studying the literature on battery research, the first step in a quest that would dominate the next eleven years of his life.

The battery project was a departure for him. For years he had railed against "storage batteries," as rechargeables were called. He saw them as catalysts for corruption, the tools of scam artists. Now he was committed to bringing the technology into a new, respectable age, and he was confident that he would succeed. "I don't think Nature would be so *unkind* as to withhold the secret of a good storage battery, if a real earnest hunt were made for it," he wrote to a friend. He had no idea what he was getting himself into.

Edison's goal was to create a new battery that would triple the capacity of the most advanced lead-acid batteries of his day. He wanted to surpass lead acid by ditching both the lead and the acid, finding new metals and electrolytes that could build a battery that was not only more energetic but also longer-lived. Part of the reason for his choice of materials was that he believed an alkaline rather than acidic electrolyte would be necessary to build a lighter and longer-lived battery. But he was also competing against the market-leading Electric Storage Battery (ESB) Company of Philadelphia, which was owned by the New York tycoon William C. Whitney, and which controlled most of the patents on lead-acid batteries. Edison couldn't chase them on their own well-established road. He would have to find a different approach.

The romantic telling of this period of Edison's life has the proudly anti-academic inventor scorning theory and, instead, systematically churning through every conceivably suitable substance—innumerable grades and forms of copper, iron, cadmium, cobalt, magnesium, nickel hydrate, along with any number of formulations of the electrolyte. As his biographer Matthew Josephson wrote, "The number of experiments mounted into the hundreds, then to the thousands; at over ten thousand, Edison said, 'they turned the register back to zero and started over again.' A year, eighteen months went by, and they had not even a clue."

In reality, he was not working blindly. He knew the literature. He was probably building on research conducted by scientists such as the Swedish chemist Waldemar Jungner, who had been doing pioneering work on alkaline batteries himself. Edison was also probably spying on his competition at ESB, which was racing to develop an improved lead-acid battery called the Exide.

Because of the intensity of the competition with ESB, almost as soon as Edison chose a basic design for his battery he began promoting it. In 1902, he wrote an article for the *North American Review* reporting that his lab work had led him to "the final perfection of the storage battery," a cell that used nickel and iron electrodes and a potassium-based electrolyte. He had his critics. In the magazine *Outing*, a writer named Ritchie G. Betts mocked Edison for promising "a featherweight and inexhaustible battery, or one which may, by the twist of a wrist or the pass of a hand, draw power, and be recharged from the skies or the atmosphere or whatnot, and lo! all problems are solved! The ideal automobile is at hand!" But the critical voices would be overwhelmed by a press infatuated with the myth of Edison, the Wizard.

By 1903, Edison's workers were dropping his nickel-iron batteries into cars and logging miles, and conducting primitive abuse testing by throwing batteries out of third-story windows of their Orange, New Jersey, lab. By the following year, they had pushed the battery to impressive new levels of capacity: 14 watt-hours per pound, 233 percent better than the lead-acid batteries of the day. It wasn't quite triple, but it was close enough.

Edison launched his Type E nickel-iron battery with a level of hype and overpromising that would do today's most egregious vaporware vendors proud. It was a "revolutionary" new battery that would "last longer than four or five automobiles." Predictably, Edison's fans in the press were enthralled. The nickel-iron battery "revolutionized the world of power." The "age of stored electricity" had arrived.

The giddiness didn't last long. Soon, the batteries began to leak. Many of them quickly lost as much as 30 percent of their capacity. And so Edison recalled the batteries he had trumpeted so loudly, went back to the lab, and set out to finish what he called his "damned problem."

Five years passed. Edison's health deteriorated. It was, according to Josephson, a "prevailingly somber period." It was a grim few years for the electric car as well. The gasoline engine was improving quickly. In 1907, Rolls-Royce released a six-cylinder gas car, and Ford launched its affordable, popular Model N in 1906. The competition for Edison's battery was growing tougher with each passing year.

One of Edison's employees solved the leakage problem with a rugged sealed container, but the performance still wasn't what they hoped. Then in 1908, they had a breakthrough. The following year, Edison wrote in a letter: "At last the battery is finished." In July 1909, he released the second-generation A cell.

This battery was a success. It was nearly indestructible and had a longer life span than competitors, which made it particularly attractive to the owners of electric-truck fleets. Yet soon after the arrival of Edison's A cell and ESB's competing product, the Ironclad-Exide, Charles Kettering invented the automatic starter for gasoline engines, and that was effectively the end of the early electric passenger car. Before long ESB began adapting its lead-acid Exides for the subordinate duty of turning over an internal combustion engine. Edison's battery found work running lamps and signals in mines, trains, and ships. In World War I, it was used for telegraphy and in submarines. For the next several decades, as the gas-powered car became an emblem of the American dream and the electric car went into a long hibernation, Edison's battery and its competitors moved into supporting roles for a petroleum-driven world.

Back in 1908, two things rescued Edison's battery. The first was the addition of nickel flake to the electrode. The second was lithium.

In a patent application filed on May 10, 1907, Edison explained

that adding two grams of lithium hydroxide to every 100 cc of electrolyte solution caused his battery's capacity to spike by 10 percent and
extended the amount of time the battery could hold a charge by a "remarkable" amount. Today we know that the lithium hydroxide most
likely helped avert some detrimental, unintended chemical reactions
that had been sapping away the battery's strength. Edison, however,
had no clue why it worked, and he probably didn't care.

Edison didn't build anything resembling a true lithium battery. Lithium was the salt in his stew. But if nothing else, it was a poetic choice:
a century later, after scientists have spent decades scouring the periodic table for better battery materials, we know that lithium is the best
possible foundation for electrochemical energy storage. The universe
hasn't given us anything better.

Lithium, which is now used for purposes as diverse as treating bipolar disorder and strengthening aircraft frames, is one of the three pri-
mordial elements, created during the first minutes after the big bang.
The lithium atoms in our laptops and cell phones are among the oldest
pieces of matter in the universe. Composed of four neutrons, three
protons, and three electrons, lithium is the third element on the periodic table, preceded only by hydrogen and helium. A metal, it is half
the density of water and, in its elemental form, too volatile to exist in
nature. Pure lithium is silvery-white and soft, like cold Camembert
cheese, and must be stored in oil to prevent it from reacting with air or
water.

Like its heavier alkali-metal cousins sodium and potassium, lithium
was first isolated in the early nineteenth century. In 1800, a Brazilian
chemist visiting a mine on the Swedish island of Utö discovered crystalline minerals he named spodumene and petalite, both of which we now
know are compounds of aluminum, silicon, and lithium. Seventeen years
later, Johan August Arfwedson, a young Swedish chemist working in the
lab of Jöns Jacob Berzelius, broke petalite down into a lithium salt, which
earned him credit as the discoverer of the element. Berzelius anointed
the new mineral, which Arfwedson was never able to isolate in its pure
form, "lithos," from the Greek for "stone."

By the mid-1800s, lithium salts were being used medicinally, first
to treat gout and, later, all manner of illnesses. Lithium therapy became

popular in the late nineteenth century because of the spread of the idea that illnesses ranging from gout to asthma to depression were caused by uric-acid imbalances, and that lithium, by dissolving uric acid, could help with them all. Soon lithium salts and lithiated beverages, products with brand names like Buffalo Lithia Springs Water, were being sold widely as curatives. A brewery in Wisconsin made Lithia Beer using spring water that was high in the mineral. The lithiated drink with the most lasting influence arrived in 1929, with the name Bib-Label Lithiated Lemon-Lime Soda. The Howdy Company of St. Louis marketed the soda, which contained lithium citrate, as a hangover cure. "It takes the ouch out of grouch," went an early slogan. Before long the company founder changed the drink's name to 7-Up Lithiated Lemon-Lime, and today, we know its delithiated progeny as 7UP. (The latest ad campaign: "Ridiculously bubbly!")

Lithiated soda might have been dubious, but it was harmless. The next major medical application of lithium was far less benign. In the 1940s, some doctors began giving heart-disease patients lithium chloride as a substitute for their usual sodium-rich salt, and the result was a number of lithium overdoses, several deaths, and a wealth of data on how much lithium it takes to kill a human being. The timing was unfortunate. In 1949, the same year news of the lithium poisoning broke, the Australian psychiatrist John Cade reported dramatic results using safe doses of lithium salts to treat mania. Yet the toxic-overdose episode gave lithium such a bad reputation that the FDA wouldn't approve lithium carbonate as a psychiatric medication until 1970.

Lithium is now one of the most effective pharmaceuticals available for treating mental illness. Mood-stabilizing drugs such as Eskalith, Lithobid, Lithonate, and Lithotabs are indispensible for regulating bipolar disorder. Scientists still aren't exactly sure how they work, but they do know that lithium affects neurotransmitters and cell signaling, and that it increases production of seratonin, the mood-elevating compound whose shortage is associated with depression. (Intriguingly, lithium also seems to stimulate brain-cell growth.) A study published in *The British Journal of Psychiatry* in 2009, which compared suicide rates and lithium levels in the drinking water of eighteen Japanese towns, found that "even very low levels of lithium in drinking water"—0.7 to 59 micrograms per liter, compared to the nearly 340 mg of elemental lithium delivered in the commonly prescribed 1,800 mg daily dose of pharmaceutical lithium carbonate—"may play a role in reducing suicide risk within the general population." In an invited commentary piece published in the same issue, a Canadian psychiatrist suggested that lithium could one day be added to drinking water, just as fluoride is added to public water supplies to prevent dental disease. Right away the theory that government eugenicists wanted to exercise mass mind control by lithiating the water supply spread across paranoiac websites.

Despite the significance of lithium as a psychiatric tool, the pharmaceutical industry absorbs only a tiny fraction of the approximately 120,000 metric tons of lithium-bearing compounds that are mined, processed, and sold each year. The largest share goes into metal alloys, ceramics, and lubricating greases, along with various rarefied applications—devices that absorb excess carbon dioxide in the air aboard spacecraft and submarines, rocket propellant, and certain types of nuclear reactors. Because we've stopped replacing the old ones, lithium no longer contributes to the manufacture of thermonuclear weapons. Isotopes of lithium did, however, trigger the largest thermonuclear device the United States ever detonated, the bomb that in the 1954 Castle Bravo test unleashed a blast twelve hundred times more powerful than what hit Hiroshima and Nagasaki, and dusted a swath of inhabited South Pacific islands with radioactive fallout.

Of all of lithium's uses, however, the one with the most profound implications for the future—the application that has already affected the lives of billions of cell-phone-, laptop-, and iPod-using people, and the one that stands to change the way we drive and to transform the way we use energy—is in batteries.

Think of electricity as a stream of electrons. The ideal tool for storing electricity squeezes the largest number of electrons into the smallest and lightest device possible. But you can't just shove loose electrons in a can. To get an electron, you have to pry it loose from an atom. In this way, every electron you get out of a battery comes with baggage in the form of protons and neutrons, both of which are more than eighteen hundred times as massive as an electron. In the lead-acid 12-volt battery under the hood of your car, each usable electron comes tethered to a hefty lead atom—82 protons and 125 neutrons in the nucleus, for a total atomic weight of 207.2. By contrast, each electron you snatch away from a lithium atom in your cell phone comes with a burden of only 3 protons and 4 neutrons; lithium has an atomic weight of 6.941, thirty times less than that of a lead atom.

A lithium atom's eagerness to shed its outer electron also means that it can be used as the basis for batteries that are more powerful and energy dense than those based on just about any other element. In essence, a battery is a high-energy chemical reaction that has been hijacked into providing useful results rather than a burst of flames. Lithium, recall, is too reactive to exist in nature in its pure form; combine the active ingredients of a lithium-ion battery's two electrodes and, under the right conditions, you have an excellent high explosive. A battery, however, frustrates these violent tendencies. By putting an electrolyte bridge between those two electrodes, a battery keeps those bomb parts at a safe distance from each other, placing an explosion in suspended animation, creating a chemical system throbbing with energy that can be redirected and exploited.

This system, used correctly, can help plug a gaping hole in our technological ecosystem—our pathetically primitive ability to store energy. As Bill Gates put it in a 2010 speech, all the batteries in the world can together store only ten minutes of our global electrical needs. In an era of grave concern about the future of energy, this is a fairly obscene weakness.

Today we power our cars almost exclusively by burning the fossilized remains of prehistoric plankton, transforming the energy that holds those hydrocarbon molecules together into energy that moves us around town. And oil has many advantages: it's powerful, versatile, and easy to store—we can simply put it in a barrel or a gas tank and let it sit. Yet oil's many consequences (environmental degradation, greenhouse-gas emissions, the enrichment of dictators and sworn enemies of civilization), combined with the fact that we will eventually run out of affordable sources, make finding alternatives an obvious imperative.

Of the alternatives, electricity is the cleanest and most flexible option. It's piped into every home in the country. Mile by mile, it's cheap compared with gasoline. It's far more feasible than hydrogen, and in almost all circumstances it's cleaner than ethanol. It can come from almost any source—natural gas, coal, nuclear, hydroelectric, solar, wind. Even when it is generated by a coal-burning power plant, it still produces less carbon dioxide per mile than a mile powered by gasoline.

The problem is, electricity is hard to store, and that's why the lithium-ion battery has attracted so much attention. It has already proved itself to be a powerful driver of modernity. Largely because of the arrival of the lithium-ion battery in the early 1990s, the cellular telephone first became ubiquitous and then transformed into a pocketable computer. Then it became a computer that connects wirelessly to the Internet. Then it became a computer, camera, MP3 player, GPS navigator, movie player, and all-around life planner and time waster, extending the reach of the information revolution into our pockets.

Now, the hope is that lithium-ion and, later, even more advanced batteries can both make electricity a viable transportation fuel and help fill the gaps in the electrical grid that are currently stifling the implementation of renewable energy sources. Already companies are building tractor-trailer size lithium-ion battery banks and hooking them up to wind and solar farms. The ability to store intermittent sources of energy like these (the sun goes down at night, the wind doesn't always blow) makes them vastly more practical and affordable as alternatives to polluting sources such as coal.

This is the kind of transformation that the scientists who laid the intellectual foundation for the rechargeable lithium battery had in mind. They were motivated by both scientific curiosity and big-picture social concerns. They began working on the vexing problem of energy storage more than four decades ago, in an age of scarcity and uncertainty much like our own.

2

FALSE START

We have only two modes—complacency and panic.
—James R. Schlesinger, first U.S. secretary of energy

Within four decades of the gas-powered car's victory over the electric vehicle, air pollution in many world cities had reached life-destroying concentrations. This wasn't all the automobile's fault. The clouds of smog that sometimes got trapped in temperature inversions over New York or London and killed anywhere from a few hundred to a few thousand people—smokestacks were largely to blame for those. But cars were a major part of the problem. In Los Angeles, where residents occasionally had to wear gas masks *indoors*, the automobile was the primary culprit. Unfettered tailpipe emissions reacted with sunlight and transmogrified into a photochemical death cloud that could hang over the city for days or weeks at a time. In 1950, when a Caltech professor identified automotive tailpipe emissions as the main source of smog, there were a half million cars in LA, which is why Los Angeles County's efforts to crack down on industrial pollution in that decade did approximately nothing to solve the problem. The number of cars continued to grow. By 1966, the 3.75 million cars in Los Angeles County produced 90 percent of the 13,730 tons of air pollution emitted each day. Certain plants—spinach, orchids—could no longer survive in LA. The problem wasn't confined to Los Angeles, however, nor to just LA and the dense urban belt between Washington, D.C., and New York City. Certain air-quality measurements in Chicago registered high enough concentrations of carbon monoxide to turn a sober driver into a gas-drunk danger. In 1966, a group of orbiting American astronauts tried and failed several times to take a snapshot of their home base in Houston because the city was too obscured by smog.

In 1961, California began requiring that new cars sold in the state come equipped with a system that would send fumes containing unburned fuel back into the engine where they would combust rather than escape through the tailpipe. It didn't make much of a difference. By January 1967, when *Time* published a cover story on air pollution called "Menace in the Skies," a California state public health official told the magazine, "It is clearly evident that between now and 1980 the gasoline-powered engine must be phased out and replaced with an electric-power package." He clarified: the state needed to "serve legal notice that after 1980 no gasoline-powered motor vehicles will be permitted to operate in California."

Before long the backlash against the internal combustion engine intensified to a level that is difficult to imagine today. That's the difference between smog—pollution that hangs in the air, visibly choking American cities—and invisible pollution by carbon dioxide, which will wreak an indeterminate amount of destruction on the planet some decades in the future. When people can't breathe, they get desperate, and by the time Congress began debating the Clean Air Act of 1970, anti-auto sentiment had grown fierce. In California, one state legislator proposed the outright banning of the internal combustion engine.

By then, the geopolitics of petroleum production had also become nightmarish. Oil-exporting countries had begun rewriting contracts, demanding bigger cuts, raising prices, and in some cases nationalizing the Western oil companies operating on their soil. The dynamic had been building for years, beginning with the 1956 Suez Crisis, in which the Egyptian leader Gamal Abdel Nasser lashed out at his nation's former occupier, Britain, by seizing control of the narrow waterway through which the majority of its Iranian oil traveled. A more recent memory as the 1970s began would have been 1967's Six-Day War, when Egypt, Jordan, and Syria tried to destroy Israel with a combination of bullets, bombs, and the "oil weapon."

Global conflict *and* pollution aside, Americans were already burning an unsustainable amount of oil. Gasoline shortages began early in 1973, the result of mismanagement rather than international emergency. In April of that year, President Nixon gave the first presidential address on energy. Late that summer, oil began selling for more than the official posted prices. "It was a decisive change, truly underlining the end of the twenty-year surplus," the oil historian Daniel Yergin wrote.

Then in October 1973, when Egypt and Syria once again attacked Israel and persuaded the oil-exporting countries of the Middle East to levy an oil embargo against the United States, a lifestyle-threatening crisis began. Before the Arab oil embargo, in October, oil was $3 a barrel (in 2010 dollars, about $15). Soon, there simply wasn't enough to go around. By February of the following year, one-fifth of the gas stations in America would run dry.

When the oil crisis hit, the interest in electric cars that had been revived by the smog plague grew frantic. The problem was that no battery technology could match the versatility and the power of the modern gasoline engine. Electric cars had to compete with sixty years of refinements to the internal combustion engine. Some of the batteries available in the early 1970s could power a small electric car for a minimally acceptable distance, provided the driver never wanted to climb a steep hill or get on the freeway. Some batteries could dump electrons quickly for a sudden boost of power, but then they were all but shot. The available technology would power only the saddest, most anemic electric cars—nothing that would impress the drivers of the day.

At that moment, however, a small international network of scientists was shaking battery science out of a long stagnation by applying the same theories and methods that had yielded the transistor and the integrated circuit. The Stanford University laboratory of Robert Huggins was the seat of this reinvigorated research, and the graduate students and postdocs who passed through it in the late 1960s and early 1970s would go on to reinvent the field.

In 1965, Huggins had gone to the Max Planck Institute in Germany to study with a professor named Carl Wagner—the first scientist to specialize in the movement of ions (charged atoms or molecules) in solids. This sounds like a parody of an overly narrow scientific subspecialty, but in fact it was a rich vein of inquiry. The realization that ions could quickly dart around inside solid materials, almost like atoms floating in a liquid, had enormous implications for battery science. Previously, battery research had assumed that the important reactions inside a battery occurred on the surface of electrodes. Picture a plate of lead dipped in an acidic electrolyte. The reactions that make that battery run happen at the surface where the liquid electrolyte touches the solid plate; the lead inside the plate is just there, adding weight. What if you could engineer reactions that happened *inside* a solid electrode? That would change things dramatically. And that's exactly what solid-state ionics, as the study of ion movement in solids is called, did. "If you can store ions inside these materials, rather than just having reactions on the surface, you have the possibility of much greater capacities," Huggins said. Huggins didn't have batteries in mind when he went to study with Wagner, but he happened to return to the United States at the moment that battery-powered cars began to seem like the solution to multiple major problems. "I came back with a whole new set of tools, ways of looking at things that I hadn't had before," he said. "And not long after I got back here, this announcement came from Ford Motor Company."

In 1967, Neil Weber and Joseph T. Kummer, researchers at Ford's Dearborn, Michigan, campus, invented a battery that was a radical departure from tradition—the inverse of everything that had come before. Unlike the 12-volt lead-acid starter batteries used in conventional cars, which immerse solid electrodes in a liquid electrolyte, Ford's new device would do the opposite: the electrodes would be liquid and the electrolyte would be solid. To be more precise, both the positive and negative electrodes (commonly called the cathode and the anode, respectively) would be molten: one made of sulfur, one made of sodium, both heated to 300°C and separated by a solid ceramic electrolyte. In a conference paper, Huggins called it a "revolutionary" approach.

Ford's unusual electrolyte—the medium that separates the positive and negative electrodes, allowing ions to move between them while preventing the transfer of electrons—fascinated researchers most of all. The cheap, ceramic form of aluminum oxide called beta-alumina had been around for several decades, but until Ford repurposed it as an electrolyte, no one had ever given it much thought. To the human eye, beta-alumina is a glistening white solid, but on a molecular level it's like a high-rise building with no stairs; sodium ions occupy each floor, but they can enter and exit only through the windows. Putting beta-alumina at the heart of a new type of battery broke a long-standing logjam. "The sodium beta-alumina was a shock to everybody," Huggins said. "This is so different from all the battery stuff that had been going on for a long time that it was really interesting."

Researchers in academic and industrial labs around the world turned their attention to beta-alumina, Huggins's group included. In Huggins's stable was a young postdoc named Michael Stanley Whittingham. Straight out of a doctoral program at Oxford, Whittingham arrived at Stanford University in 1968, a year after the Ford announcement. At Oxford, he did his master's thesis on materials called tungsten bronzes, which conducted both ions and electrons and seemed, among other things, like promising catalysts for turning coal to gas. Shortly after his arrival in Palo Alto, Whittingham's group decided to try to find out exactly how quickly ions could move through Ford's beta-alumina electrolyte. To do so, they needed to make an electrochemical cell, and in this case that would require a particular kind of electrode material.

Whittingham's bronzes would work perfectly, because they were insertion compounds, or what in later years would come to be called intercalation compounds. "Intercalation" traditionally refers to the insertion of an extra day into the middle of a calendar—the addition of February 29 to a leap year, for example. In this context, "intercalation" describes a class of crystalline materials that ions can be *inserted* into without changing their underlying structures. On a molecular level, these bronzes were filled with tunnels, and in the right kinds of chemical reactions, ions can be "inserted" into these tunnels and then yanked out again, repeatedly, without altering the structure of the insertion compound itself. Those experiments were strictly academic, but they were essential for building the knowledge that would soon deliver the world's first rechargeable lithium battery. "Things were exciting. Things were going on," Huggins said. "The integration of solid-state electrochemistry into areas of application like batteries and fuel cells—that was brand-new." The work was fundamental, but the people in Huggins's lab had idealistic goals. As Michel Armand, one of Huggins's graduate students from that era, said, "I bought an old car, which was of course a whale on wheels, making something like four miles per gallon in the city. I was from this time convinced that we had to do something with transportation."

By 1972, enough international scientists were working on solid-state ionics that it was time for a conference. That September, Huggins, Whittingham, Armand, and eighty others gathered in the alpine village of Belgirate, Italy, a hamlet in the mountains north of Milan, where they shared ideas on putting ion transport to use building batteries and fuel cells. Most of the attendees were in their thirties and forties, but there was also an old eminence among them: Carl Wagner. "We were all very pleased he was there," Huggins said.

Gathered a little more than an hour's drive from Alessandro Volta's hometown of Como, the Belgirate delegates talked about every exotic battery chemistry they could imagine in those days: sodium sulfur, lithium sulfur, lithium aluminum iron sulfide, zinc bromine, lithium chlorine. They discussed the feasibility of magnesium oxygen, sodium oxygen, lithium copper fluoride, and zinc silver dioxide. They lusted after the most theoretically promising candidates of all: the "metal-air" batteries— zinc-air, magnesium-air, aluminum-air, sodium-air.

It was a foundational meeting, the beginning of a narrow subsubdiscipline that would have an outsize influence on the world. A yellowed black-and-white photo from the conference proceedings shows the group posing before a stand of conifers, a class picture of a scientific community that didn't yet realize it existed. Today the living members of that delegation are the dons of the academic battery-research scene, the old-timers who after forty often frustrating years have finally seen their work vindicated.

By the time of the Belgirate conference, industrial research into electric drive and advanced batteries was expanding rapidly. In 1972, GM, Ford, Chrysler, and American Motors were all working on electric cars. So was Toyota. So was a coalition of eight German companies that included Daimler-Benz, VW, Bosch, and Siemens. So was Fiat. So were national efforts in Japan, France, and England. As for batteries themselves, in addition to university-based programs, scientists at Argonne National Laboratory, Bell Labs, the Electric Power Research Institute, Dow Chemical, and General Electric were all scouring the periodic table for the solution to the battery problem.

Oil companies were at it too, including the largest of the so-called majors: Exxon. The oil giant believed that in a few decades, most likely after the turn of the millennium, petroleum production would peak, and that the time to diversify was now. They did so by starting a division called Exxon Enterprises, which operated like a venture capital firm. With the backing of the richest industrial company on the planet, Exxon Enterprises attempted to break into businesses as diverse as office equipment, nuclear reactors, and solar panels.

Then as now, Exxon prided itself on being run by engineers, and for the new venture they raided all the best schools, hiring the brightest technical minds they could find and assigning them to basic research that could be applied to any number of new inventions. Michael Stanley Whittingham was one of them.

Just after the Belgirate conference, Exxon Research and Engineering lured Whittingham to the grim industrial corridor of eastern New Jersey. With practically unlimited research funds, his job was to conduct fundamental research on everything energy related except oil. In a small laboratory in Linden, with a refinery staring at them from across Routes 1 and 9, he and his colleagues were to perform the research that would keep Exxon in the black once the pipes across the street were empty.

First, Whittingham and a few colleagues went to work on superconductors, the idea being that if you could find a material that conducted electricity with no resistance at room temperature, then (theoretically) you could dramatically increase the efficiency of any electrical system, not to mention build an entirely new generation of electronics. They started by injecting ions into tantalum disulfide (TaS_2), which at the atomic level is like a crystalline sandwich, with an empty spot (called a "galley") in the middle where ions could go. Sometimes, those ions could make tantalum disulfide do interesting things. Normally the material became a superconductor at 0.8° above absolute zero. With potassium ions inserted into those galleys, however, that temperature increased significantly.

Whittingham began treating various materials with potassium hydroxide, trying to understand why adding potassium ions to tantalum disulfide raised its superconducting temperature. In the process, he noticed that TaS_2 injected with potassium had an extremely high "free energy of formation"—that each molecule had a lot of energy tied up in its chemical bonds. Soon he and his colleagues had an idea: "We said, 'Hey, we can store energy in this,'" Whittingham said. When Whittingham's team told their superiors that they might have the raw materials for a new, powerful battery, the managers at Exxon immediately jumped to the idea of an electric car.

Whittingham's group realized that tantalum was too heavy to go into a battery, so they decided to replace it with the lightest transition metal, titanium. Soon they were experimenting with titanium disulfide (TiS_2), another molecular sandwich structure. Paired with the right negative electrode, titanium disulfide could make a battery with a theoretical energy density of up to 480 watt-hours per kilogram, more than twice what was generally accepted as necessary to power a viable electric car. And titanium was an ideal ingredient—it was light, abundant, and an excellent conductor of electricity. They initially tried pairing TiS_2 with a negative electrode made of potassium, but potassium was extremely hazardous to handle. Instead, Whittingham turned to lithium.

Whittingham said that lithium came to mind because Japanese fishermen had recently begun using lithium-based primary (nonrechargeable) batteries on fishing floats so they could see their nets at night. Still, the idea of a rechargeable lithium battery had been in the air for a while. It had come up at the Belgirate conference, and Sohio (another oil company), General Motors, and Argonne National Lab were all working on lithium-based batteries around the same time. The difference was that all of those projects involved extremely high temperatures—designs similar to that of Ford's sodium-sulfur battery, which used molten electrodes and as a result had to be kept impractically hot.

It didn't take long for Whittingham and Exxon to realize the promise of what they had created. Of all the competing chemistries available in those years, theirs was the only lithium-based compound that worked at room temperature. And so when Whittingham crossed the Hudson into Manhattan and presented his work on lithium batteries to a committee of Exxon board members in the company's fortresslike Sixth Avenue headquarters, it was an easy sell. At the time Exxon was eagerly expanding into alternative businesses. The technology seemed like a breakthrough. The project fit perfectly with Exxon's desire to move into electronics and alternative energy. The answer came quickly: let's put some money into it.

Manufacturing Whittingham's battery fell to a man named Bob Hamlen, a self-professed battery geek. Before the call came from Exxon, he was head of electrochemistry at General Electric, and he experimented with batteries in his spare time. In 1973 he moved from upstate New York back to his home state of New Jersey and began reporting each day to Linden, where he set about scaling Whittingham's creation into a meaningful business.

The biggest question when Hamlen arrived was what to use for an electrolyte. In batteries that operate at room temperature, most electrolytes are a solution of two materials: a liquid (the solvent) and a salt (the solute). For the battery to work in cold climates, the liquid must have an extremely low freezing point—in the neighborhood of 30°C below zero or, if possible, even lower. It must be an electrical insulator (something that doesn't conduct electricity) to keep the ionic reaction and the electronic reactions separate, to force electrons up and out of the battery. Finally, it has to dissolve a salt that breaks down into the right kind of ions needed for the electrochemical reaction, so the salt, which Hamlen's group also had to find, had to match the solvent perfectly.

Hamlen's group started by dissolving lithium perchlorate in dioxolane, a clear, combustible organic liquid. When lithium perchlorate dissolves positively charged lithium ions break away from the negative ions—in this case, beautifully symmetrical clusters of chlorine surrounded by four oxygen atoms. If the temperature spikes because of a short circuit, all that oxygen can react with the hydrogen and carbon in the solvent. This fact, along with the inherent volativity of the metallic lithium anodes used in Whittingham's battery, kept the lab work interesting. After several visits to the Linden lab, the fire department threatened to make Exxon pay for the special chemicals required to put out lithium fires.

Soon, a researcher on Hamlen's team developed a solute that worked well enough that the company decided to show off its progress. Hamlen's group built several test batteries and sent them to a Society of Automotive Engineers conference in Chicago. "You can't ship lithium on a plane, so we sent a guy on the train to take 'em out there," he said. While their colleague was in Chicago putting on a show, Hamlen and a colleague discovered something unfortunate: their new electrolyte was slowly decomposing, emitting gas that was almost certainly building pressure inside the same batteries that the members of the Society of Automotive Engineers were supposed to be marveling over. Worse, the gas that was bubbling up inside was diborane, which bursts into flames upon contact with moist air.

"I think back at some of the dumb things you do," Hamlen said. "We called and said, 'Take the cells back to your room each evening.'" Fortunately, they had installed a small vent on the top of each cell. "'Carefully unscrew the vent a little bit until the gas pressure gets relieved. But get your hands out of the way, because it's going to catch fire as soon as it comes out.'" And so each day, Exxon's man in Chicago would show off the company's breakthrough rechargeable lithium batteries. Then each night, back in his hotel room, he would carefully twist the top off each battery and watch as a fireball leaped out.

The next electrolyte they tried was safer, but there was a tradeoff: the calmer electrolyte made for a less powerful battery. They had also replaced the volatile metallic-lithium anode with aluminum, making the battery safer yet. The group had made enough progress that by 1976 it was time to go public. That year Whittingham published a landmark paper on the $LiTiS_2$ battery in *Science*. Exxon opened a developmental facility in Branchburg, a small town thirty-some miles west of the industrial squalor of East Jersey. Their headquarters was the first occupant of the freshly bulldozed Branchburg Industrial Park. Exxon, king of oil, was in the battery business.

By 1976, the battery industry seemed to be on the verge of a boom. The previous year *Forbes* had declared the battery business, "of all things," to be "one of today's hottest items." In 1976, Congress passed the Electric and Hybrid Vehicle Research, Development and Demonstration Act, which aimed to stimulate the production of serious alternatives to the gasoline engine. Interest in electric cars and the batteries to power them was so urgent that Congress passed the bill over President Gerald Ford's veto.

In October 1976, an article in *Forbes* declared that "despite present—and formidable—problems, the electric car's rebirth is as sure as the need to end our dependence on imported oil." Industry publications were brimming with confidence too. According to *Chemical Week*, "After a hiatus of almost 50 years, electric vehicles are poised for a comeback . . . And this time, electric vehicles have a reasonable chance of forging a competitive niche in both commercial and passenger vehicle areas."

Exxon began talking loudly about its fears for the future of oil and its hope for new sources of energy. The oil industry was so badly disrupted that some journalists began speculating about Exxon's ability to even survive. As *Forbes* put it, "Given two major trends, one geological and the other political and social, the mighty Exxon Corp. could be forced into at least partial liquidation within a decade . . . It probably won't happen. But it could." The primary reason was that oil seemed to be running out, quickly. "Unless the presently unexpected occurs, the world's petroleum reserves are within a few years of their peak and will begin a slow decline to the point where oil and gas will be too valuable to use as energy," the article continued. Instead, the world would have to use what oil was left in the ground for other petroleum products and find something else to power its cars.

George Piercy, the Exxon executive who ultimately oversaw Exxon Enterprises, was the leading delegate for the major oil companies during the disastrous nonnegotiations that preceded the first oil crisis. In a Vienna hotel room in 1973, Piercy was the one to tell Sheikh Yamani, the Saudi oil minister, that the oil companies refused to pay the 100 percent increase OPEC was demanding. Piercy said that he simply did not have the authority to agree to OPEC's demands. A price increase that steep would disrupt the economies of the consuming countries so greatly that he would have to consult with those governments before making any deal. Yamani picked up the phone and called his colleagues in Baghdad. He hung up, turned to Piercy, and said, "They're mad at you." When Piercy asked Yamani what came next, Yamani famously replied, "Listen to the radio." A little over a week later, the Arab oil embargo began.

Piercy was therefore better aware of the precariousness of the oil companies' position than perhaps any of his contemporaries. He recognized the need for alternatives, hence the company's interest in batteries and motors for electric cars. Still, when the time came to start selling Whittingham's battery, Exxon had to start small. Very small. Their debut product—the first rechargeable lithium battery ever to reach the market—was a duo of button-size cells intended to run a solar-powered "Perpetual Watch" that the Swiss company Ebauches (now part of the Swatch group) wanted to build. The battery division published a pamphlet aimed at commercial customers introducing their breakthrough battery. "This may look like an old familiar button cell battery," read the text beside a coin-size silver disc. "It isn't." No, this was the result of "new advanced technologies in energy storage," a novel approach that "provides one way for man to store the diffuse and intermittent light that reaches him from the sun." Exxon was no longer just an oil company, the message went. "In a time of growing awareness of energy resources and needs, Battery Division is concerned exclusively with superior energy storage technology."

The watch might not seem like an obvious first application for Exxon's battery, but the arrival of the digital watch in Japan in the 1970s was in fact a subtle but important pivot point for battery technology— the moment batteries began to change from something that you kept in a drawer to something you carried around on your person. The digital watch was also the first widespread application for tiny lithium primary batteries. "The digital watch really brought the wearable battery, if you like, to the mass market," said Peter Bruce, a longtime lithium battery researcher at the University of St. Andrews in Scotland. "And I don't think it's a huge leap of imagination to envision other devices that require power that you might be carrying around."

Exxon's battery had never powered anything larger than some alarm clocks that the company used as promotional devices, but Whittingham, Hamlen, and the other true believers understood that this was the natural course of things. The technology would eventually scale up. And in the small-device market, Exxon's battery had a few major advantages over its competitors. Nickel-cadmium batteries bled away their energy quickly; silver-zinc batteries died after being charged and discharged only twenty to twenty-five times. Exxon's tiny, hermetically sealed cell had a higher voltage than its competitors and an intrinsic flexibility that meant that, in the "solar watch" that was the goal of many watch designers of the day, it could provide "the opportunity for indefinite watch operation without battery replacement." As long as a solar cell kept charging the battery, it would essentially never die.

While Hamlen was working to scale up the watch-battery business, he evangelized for the long-term promise of the Exxon Compound, declaring it the basis for the most promising electric-vehicle battery yet. In a presentation at the 1978 meeting of the American Association for the Advancement of Science, he said, as paraphrased by the trade publication *Chemical Week*, that the battery "may be the most desirable power source for future electric autos from the standpoint of cost and efficiency," and that "projected performance levels of the battery should make it possible to build a two-passenger electric vehicle, with an urban driving range of 100 miles, at a cost of approximately $5,000."

Exxon was moving into the electric-car business on other fronts as well. In 1979, the company spent $1.2 billion to buy Reliance Electric, a manufacturer of electric motors in Cleveland. An electrical engineer at Exxon Enterprises had created what the company called the alternating-current synthesizer (ACS), a controller for AC electric motors that enabled one to vary the speed of the motor for maximum efficiency. Exxon made bold claims for the ACS and used it as justification for acquiring Reliance, which the Department of Justice, in trust-busting mode, wanted to prevent. ACS, Exxon argued, could eventually become standard equipment on the millions of electric motors that run industrial fuel pumps, compressors, fans, and blowers. As *The Economist* put it, "What Exxon is saying is that, if half the industrial motors in the 1–200 horsepower class in America used its new type of controller, by 1990, that would save the country the energy equivalent of 1m barrel of oil a day." In other words: "The largest of the seven oil majors is gearing up for the day when oil begins to run out."

They were indeed. The problem was simple: "We're not finding as much oil as the world is using," Exxon's chairman, Clifton Garvin, told *BusinessWeek* in July 1979. "In the long term, I'd say that you don't ignore any source of energy. We can't go back to the complacency of two years ago."

Exxon in those days had a soft spot for synthetic hydrocarbon fuels— shale oil, gasified coal, and the like—but Garvin made it a point to emphasize the inevitable importance of the electric car. Exxon had no desire to build electric cars itself, he said, but through Reliance, he hoped to supply the motors, and through the Battery Division, the power. "I happen to believe that somewhere down the road, in 30 or 40 years, we're going to be fundamentally an electrically based society," he said, "and we're all going to be tooling around in electrical cars."

By October 1979, the electric car seemed to be on the cusp. *Fortune* ran an upbeat piece pegged to developments at Exxon and GM headlined "Here Come the Electrics." GM had announced a new battery, a zinc-nickel-oxide power pack that it was putting in a car called the Electro-vette—a Chevette with a backseat full of batteries and a hundred-mile range (provided you didn't drive faster than 50 mph). Exxon had by then built a prototype hybrid gas-electric car, a converted Chrysler Cordoba. The company reiterated its desire to sell motors, not build electric cars, and as *Fortune* snidely noted, "That may be just as well. It is not in G.M.'s league in marketing savvy." Whereas GM made a point of sexing up its science project by calling it the Electrovette, "Exxon refers to its car as the 'prototype, hybrid electric vehicle.'"

Faster than it came together, the electric-car surge fell apart.

First came the recession of 1979–1980, which sent Exxon and everyone else into cost-cutting mode. Unprofitable expansions into solar panels and batteries quickly came to be regarded as unaffordable diversions from the core objective of the day, which was survival. "It was a period of turmoil within Exxon Enterprises," Bob Hamlen said. "Eventually they came to the conclusion that they only wanted to get into products that had the potential to be a billion-dollar business, and if you come right down to it nothing will meet that criteria." Nothing, that is, except oil.

Hamlen's team was still conducting tests with Ebauches when he attended a meeting that sealed the division's fate. "After one presentation where we said, 'Something like this could be a neat $50 million business, but it's tough to make it a billion,' they said, 'Hell, if that's the case we don't want it. We'll sell it off and license it out.' Which is exactly what they did."

Hamlen's job was now to dismantle the Battery Division. Exxon licensed Whittingham's technology to three companies: one in Japan, one in Europe, and one in America. The American company was Eveready, at the time owned by Union Carbide. Eveready engineers suddenly found themselves in possession of boxes filled with Whittingham's data.

If recession and a slump in oil sales painted a target on the various ventures of Exxon Enterprises, the subsequent oil glut buried them for good. By 1986, oil was again below $15 a barrel, and supplies appeared to be steady as far ahead as anyone was willing to think. Governments and oil companies hadn't just started building batteries and solar cells during the oil shock. They also scoured the rest of the planet looking for more petroleum, and they found it—major reserves in the North Sea, Alaska, and Mexico. Britain, which was nearly choked to death by the crisis in the Suez, had become an oil-*exporting* country. Moreover, the conservation efforts put in place in America in the mid-1970s worked spectacularly; the American Corporate Average Fuel Economy (CAFE) requirements, which set average gas mileage standards at 27.5 mpg, saved two million barrels of oil a day between 1975 and 1985. And because oil companies had hoarded petroleum throughout the crises of the 1970s, there was more than enough to go around. New exploration, conservation, and hoarding all conspired to ensure that by the mid-1980s, national labs and major corporations had lost all interest in developing alternatives to oil.

The repercussions of Exxon's decision reached far beyond that one company. "When Exxon stopped, the federal government in their ignorance decided, 'If Exxon's not doing it, it's not worth doing,'" Whittingham said. "Other companies did similarly. Instead of saying, 'Well, why did Exxon stop? Here are the technical issues, we should now take time to address them.'"

The election of Ronald Reagan in 1980 put a temporary end to government interest in alternative energy. "If Reagan had continued the programs the Jimmy Carter administration started, we'd be a lot further ahead," Huggins said. "But that didn't happen, and so we had this hiatus."

The lack of industrial and governmental funding put advanced battery research on hold. "As money disappears, professors do something else," Huggins said. "If you want to support your graduate students, you have to get money. And money comes in the U.S. mostly from government, so what the government's interested in has an immense amount of influence over what goes on. You see examples of this all over the place. If you look at electrical engineering, you'd find there was tremendous activity on lasers for many years. Why? The military was interested in lasers. People go where the money is."

"Reagan came in and cut back energy efficiency and renewable energy programs by something like eighty percent," Elton Cairns, who at the time was working on advanced battery research at Lawrence Berkeley National Laboratory, told me. "All labs, including ours, suffered layoffs as a result. The reduction in funding occurred something like overnight. That pretty well put an end to the significant involvement in DOE labs in battery and fuel-cell programs at that time."

By the time the false start of the 1970s came to an end, the intellectual advances of those urgent years had failed to translate into commercial breakthroughs; the best batteries on the market that year could store 30–35 watt-hours/kg, making them five hundred times less energy-dense than gasoline.

At Exxon, Clifton Garvin drifted into the complacency he had warned against eight years earlier. "We're not interested in being in businesses long-term that don't meet the kinds of return criteria we see in oil and gas," he told *Fortune* in 1984. That same year, General Electric canceled its research into sodium-sulfur batteries. "Without a market, what's the sense of development?" a GE researcher told *Chemical Week*. At the time, Elton Cairns explained the dynamics of the battery business to *Chemical Week* in a simple formula that remains true to this day: The key to the feasibility of advanced batteries of all types is the price and supply of oil.

Decades later, after the lithium-ion battery had put GPS-enabled, satellite-linked computers in every middle-class pocket, had begun to make the long-standing dream of an electric car a reality, and had been widely cited as the key enabler of a clean-energy future, Bob Hamlen told the story of his time at Exxon with an audible tinge of regret. "Now there was one thing," Hamlen said. "I talked to Stan [Whittingham]— we looked back and we think to ourselves: Why on earth did we never mix the lithium with carbon? Which, as you know, is what made the current lithium-ion batteries feasible. Well, that's easier said than done because it takes particular kinds of carbon, and it might not have been so easy to identify without some past experience . . .

"I have my own personal saying, which is that quantum jumps are only in the eyes of the uninvolved. If you look carefully at every advance that's ever been made that gets reported as a big leap, you find that people involved have been doing this for a long time. I know a lot about it."

3

THE WIRELESS REVOLUTION

After the failure of Exxon's battery venture, only an entirely new approach to the rechargeable lithium puzzle could revive the technology. It would come out of the laboratory of John Bannister Goodenough, a physicist turned solid-state chemist who in the two decades after the demise of Whittingham's battery would help invent the three major strains of lithium-ion battery in use today, including the one that made the gadget boom of the 1990s and 2000s possible.

Goodenough is now a professor at the University of Texas in Austin. Eighty-eight years old and, technically, long retired, he still conducts research and tends to a flock of graduate students. He is tall, but these days his height is mitigated by a hunch; when he walks, he relies on a four-footed metal cane. Regardless, he has a commanding presence, with feathery, owllike eyebrows and a booming laugh that erupts unexpectedly and then lingers just long enough to make you uncomfortable. He has a grandfatherly demeanor and likes to offer unsolicited lessons on the Meaning of It All: "Don't forget your roots," he told me. And later: "Life goes by so quickly."

Goodenough was born in Germany in 1922, to American parents. He spent the second eleven years of his life in a small town outside

New Haven, Connecticut, where his father was an assistant professor of history at Yale. He recalls his childhood with a mixture of nostalgia and deep black melancholy. The nostalgia is for his dog Mack, the wisteria vine on the side of the house, the apple trees, the windmill in the backyard. Yet he writes of a "deep hurt that plagued my early years," in his slim, self-published spiritual biography, *Witness to Grace*. His parents were college-town gadabouts. His closest friend as a kid was Mack. He struggled to learn to read, and so he finds it miraculous that at the age of twelve he was accepted to Groton, the prestigious Massachusetts boarding school.

At Groton he came into his own. "I was never homesick," he wrote. "I was happy my parents did not come to visit to intrude on my new life." After Groton he enrolled in Yale, with the goal of finishing as much college as possible before being called to war, which, in 1938, seemed inevitable. He took a freshman chemistry class to satisfy his science requirement and to keep medical school an option. He wrestled with the responsibility of a young man living in an insane world. "I began to understand that any meaning to a life is not the accolade of others, but the significance and permanence of what we serve," he wrote.

What seemed most permanent to him was science, and he decided to begin with the fundamentals, studying philosophy of science and physics. As he approached graduation, his mathematics professor encouraged him to join the Army Air Corps as a meteorologist rather than signing up to be a war hero in the marines. Before long, Goodenough was dispatching aircraft across the Atlantic, first from a station in Newfoundland and then from the Azores. He made it through the war without major incident and remained stationed in the Azores until, in 1946, a telegram arrived instructing him to report back to Washington. Back in the States, someone had happened upon an unspent sum of government money and applied it to "reintegrating a few promising scholars to civilian life." Thanks to the recommendation of one of his professors at Yale, Goodenough became one of twenty-one officers chosen to study either physics or mathematics at Northwestern University or the University of Chicago.

Before leaving for the war, Goodenough had spent time reading Alfred North Whitehead's *Science and the Modern World*, and he made a decision: "If I were ever to come back from the war and if I were to have the opportunity to go back to graduate school, I should study physics." He hadn't thought about this in years, but the new opportunity seemed like an omen. He enrolled in the graduate physics program at the University of Chicago, where he learned quantum mechanics from Enrico Fermi. He wrote his dissertation on solid-state physics, measuring the "internal friction of iron wires doped with carbon or nitrogen." Upon enrollment years earlier, he had ignored the registrar's words of encouragement: "I don't understand you veterans. Don't you know that anyone who has ever done anything interesting in physics had already done it by the time he was your age?"

Goodenough would go on to do plenty of interesting things, just not, strictly speaking, in the field of physics. After completing his Ph.D., he took a job at MIT's Lincoln Laboratory, the federally funded defense lab that was home to several early milestones in computing, including the invention of TX-0, the first fully transistor-driven computer. At Lincoln, Goodenough found himself involved in research on random-access memory; his group's assignment was to develop a magnetic material that could store data for the gymnasium-size vacuum-tube computers of the era. He quickly found a niche as an interdisciplinary fixer, a liaison between engineers, chemists, and physicists, all of whom were working toward the same goal but none of whom spoke the same language. He developed an intuitive understanding of the interatomic terrain where chemistry, physics, and engineering collided.

Around the same time Goodenough arrived at Lincoln, two scientists at another famous lab a couple hundred miles down the eastern seaboard were inventing the cellular telephone. It was 1947. The Bell Laboratories researchers were named Douglas H. Ring and William Roe Young. Their idea was to spread a large a number of low-power base stations—essentially, radio transmitters and receivers—over a large geographical area, creating "cells" of radio coverage. When a user moves from one cell to the other, the signal is handed off from base station to base station, an innovation that would allow the system to reuse frequencies, thereby increasing the number of signals that the system can carry.

Nothing much came of the idea until the 1970s, when additional decades of research at Bell Labs, along with developments such as the commercialization of microprocessors and the advent of computerized switching stations that could control the handing off of calls between cells, begat the earliest experimental workable mobile phones, which two people could use to carry on a normal conversation without pushing buttons and saying "over and out."

As the wireless telephone became more feasible, a race began between Motorola, king of the two-way radio, and AT&T, which in those days controlled 80 percent of the American phone market—a rivalry that would lead directly to the birth of the handheld mobile phone. On April 3, 1973, a Motorola researcher named Martin Cooper claimed victory when he made the first handheld mobile phone call while walking around the streets of New York City. Using his two-and-a-half-pound phone, he rang his rival at Bell Labs to let him know who had won.

On April 3, 1973, John Goodenough was still at Lincoln Lab, and though he wasn't thinking about cell phones, he was getting restless. He was looking for a new way to apply his fundamental work in solid-state science to projects of societal importance. And in 1973, the most urgent project for a guy with his training was looking for new sources of energy. "It was obvious already in 1970 that our dependence on foreign oil was making the country as vulnerable as the threat of ballistic missiles from Russia," Goodenough writes in his memoir. The only alternative energy sources that made sense to him were hydropower, geothermal, solar, wind, and nuclear. Hydropower was already widespread; geothermal was too geographically limited; he wasn't a nuclear scientist. Process of elimination brought him to solar power. But he soon realized that solar power had the same problem as electric cars—no good way to store electricity—and so he turned his attention to energy storage.

At Lincoln, he worked on photoelectrolysis, fuel cells, and sodium-sulfur batteries, which he had been introduced to some years before when the Department of Energy asked him to help assess the potential of the Ford Motor Company's 1967 breakthrough. Then in 1974, frustrated with the growing bureaucracy involved in government-funded research, he decided to leave Lincoln. He flirted with an opportunity to found a solar research institute in Iran using a $7 million grant from the shah. Before he committed to the Iran project, however, he was invited to make a less drastic move. A letter arrived from Oxford University encouraging him to try for an open position as the head of the university's inorganic chemistry lab. He thought he had almost no chance. He was a physicist, after all; he had taken only two chemistry courses in his life. Somehow, though, he got the job.

When he arrived at Oxford in 1976, he turned to the energy-storage research that would dominate the rest of his career. He investigated methanol-air fuel cells and electrolysis before he turned to the design of new materials for lithium batteries. He knew about the titanium-based battery that Stanley Whittingham had developed at Exxon. He was also aware of its limitations. For safety reasons that the Linden, New Jersey, fire department would have understood, he decided against using a metallic lithium anode. Whittingham's group had solved the combusting battery problem by switching to aluminum anodes, but Goodenough felt it necessary to move to an entirely new battery design.

Goodenough knew that a rechargeable lithium battery based on oxides (rather than sulfides such as the one Whittingham used) would reach a higher voltage. Deciding which oxide to focus on was difficult, but in 1978 he got a clue. An undergraduate—"I can't remember the name of the girl, and I feel badly about that," he said—wrote a thesis on the structure of the $LiMO_2$ oxide, where M is a variable standing for any number of transition metals. It reminded Goodenough of research he had done back in the 1950s on lithium nickel oxides. He wondered how much lithium could be removed from it before it started to crumble. A Japanese physicist named Koichi Mizushima happened to be visiting from the University of Tokyo; Goodenough paired Mizushima with his postdoc, Philip Wiseman, a chemist, and gave them an assignment: Make variations on this compound using chromium, cobalt, and nickel, then see how much lithium you can take out before it becomes unstable. See what kind of voltage each substance would deliver in a practical lithium battery.

Mizushima and Wiseman pulled half of the lithium out, and two of the candidates were stable: good. And the voltage? An excellent 4 volts, a leap beyond the practical 2.4 volts that Whittingham's battery delivered.

One of the keys to Goodenough's breakthrough was that he turned the accepted logic about batteries—that you had to build them fully charged—on its head. By building a battery that began its life in the discharged state, he could use compounds that were completely stable in ambient air (that is, that didn't have to be built in a dry room or an argon chamber) and get a higher voltage than was possible using the standard method of building a battery full of juice from the beginning. Yet Goodenough couldn't find a single battery company in England, Europe, or America that wanted to license his invention. Not for any good reason. "Because it was unorthodox," he said.

This reversal of orthodoxy is not, in fact, a problem. Today it's the way the billions of batteries that power nearly all the world's wireless gadgets are built. Which is why Goodenough likes to conclude the story of the lithium-cobalt-oxide cathode, the innovation that was essential for getting the lithium-ion battery into the marketplace, with this: "And that started the wireless revolution."

Between Martin Cooper's first cell-phone call in 1973 and Goodenough's publication on lithium cobalt oxide in 1980, the cellular telephone had gone nowhere. Part of the problem was that the FCC took about a decade to regulate the frequencies the companies would use to carry calls. The major phone companies, for their part, thought that hardly anyone even wanted mobile phones, and certainly not enough people to justify building an entirely new infrastructure.

But that was just in the United States. By the end of the 1970s and the first years of the 1980s, small cell-phone networks were going up in Japan, Bahrain, Saudi Arabia, and all of the Scandinavian countries. In August 1981, Mexico City got North America's first network. On August 24, 1982, the U.S. telecom landscape changed dramatically when the Justice Department broke up AT&T and established a rule that each market had to have two carriers. One of those carriers would be the local landline phone company. The other could be just about any entrepreneur who won the lottery in which the FCC was raffling off licenses to chunks of spectrum.

Finally, on October 13, 1983, in Chicago, the newly created Baby Bell called Ameritech Mobile Communications launched the Advanced Mobile Phone Service (AMPS), the first commercial cellular network. From a car outside Soldier Field, Ameritech's president, Robert Barnett, made the first call, to Alexander Graham Bell's grandson in Berlin. *The New York Times* called AMPS the "nation's first commercial cellular mobile radio service, a computerized telephone technology that supporters say will vastly expand the capacity, and improve the quality of, car telephone service." Some cities already had car-phone networks, but those systems, called Improved Mobile Telephone Service, were crude relics based around a single high-power antenna that had access to only twelve channels. That meant that an antenna could support only twelve car-phone calls simultaneously. AMPS used the scheme that the Bell Labs researchers had envisioned three and a half decades earlier: an array of small base stations, low enough in power that they don't interfere with one another, covers a series of geographical "cells"; as a user travels from cell to cell, the base stations, controlled by a computerized switching system, assign that moving call a new frequency as soon as it enters a new cell, with no perceptible disruption in the conversation. Good thing, because at approximately $3,000 for an Ameritech car phone, a $50-a-month service fee, and rates ranging from 24 to 40 cents a minute, these calls were too expensive to be interrupted.

In 1984, the first handheld mobile phone went on sale—the production version of Motorola's shoe-box-size DynaTAC 8000X, which ten years after being first unveiled in prototype form could be yours for only $3,995. The FCC had approved the phone only the year before, the same year Motorola first deployed its DynaTAC cellular network in Baltimore and Washington, D.C. The DynaTAC eventually became an ironic cultural icon thanks to the 1987 movie *Wall Street*. Michael Douglas, strolling along the beach in the Hamptons, talking into a cell phone the size of a toaster? That was the Motorola DynaTAC, the first of its kind.

Just as those first enormous phones went on sale, batteries began to make news again, though Goodenough's lithium-cobalt technology still hadn't found a taker. The early 1980s gadget boom, led by Sony's

Walkman, drove a major spike in consumption of disposable alkali batteries. The Walkman in particular was a power hog, because its batteries had to power a motor that turned the wheels of cassette tapes. The Walkman's popularity was a gift to the battery industry. "We are really an adjunct of the microelectronics revolution," the president of Rayovac told *Forbes* in 1982.

The Walkman and the gadgets it inspired were not, however, good for Japanese landfills. In the early 1980s in gadget-obsessed Japan, pollution from batteries became an urgent, high-profile environmental issue. A *New York Times* piece from 1984 makes it sound as if Japan was drowning in mercury-laden disposable batteries: "Mercury, a toxic metal used in most batteries, is starting to seep into the soil around garbage dumps . . . The leakage has raised fears that Japan is slowly being contaminated by the dry cells that power its calculators, cameras, portable stereos and watches." Mercury contamination wasn't the only problem the Japanese had been having with batteries—there was also the "swallowing issue." In growing numbers, infants had been swallowing ever-shrinking batteries, and, as the *Times* put it, "a swallowed battery can burn holes in the intestines and cause inflammations." The director of Japan's Poison Control Center told the newspaper that they suspected that there were some "5,000 cases of battery ingestion in Japan each year." Together, the two issues caused serious problems for the battery industry. "Usually, nobody's interested in batteries, but suddenly we've become very famous," a spokesperson for the Japan Battery and Appliance Industries Association said.

The Japanese quest for new, nontoxic battery technology, combined with Japanese mastery of the consumer-electronics industry, began a reorientation of the battery industry, one that assured that when rechargeable batteries one day became vitally important—when all this was a matter of energy security rather than digital watches and fancier cell phones—Japanese companies would dominate. At the beginning of the 1980s, the American company Union Carbide, which owned Eveready, stood astride the world battery market. Globally, Panasonic was a powerhouse, but in the United States, Panasonic lagged behind the American players Duracell and Rayovac. Yet while Japanese companies spent the 1980s scrambling to develop the batteries of the future, the American companies largely stood still.

"At that time in the West there was really no interest in lithium batteries," said Peter Bruce, who worked as a postdoc in Goodenough's Oxford lab during the 1980s. The hot topic in solid-state chemistry in the 1980s was high-temperature superconductivity. With the energy crisis of the 1970s in the past, batteries were boring, and superconductivity—which was being portrayed in IBM television commercials of those years by a scientist in a lab coat levitating a piece of metal over a supercooled disc—seemed like magic. Both involved solid-state ionics, which made Bruce perfectly qualified to start a career in the sexiest science of the time. "That would have been the easy option, because there was a lot of funding around, and there was no funding in lithium batteries," he said. But then he went to Japan for a month as a visiting professor, and while he was there he happened to visit a couple of electronics companies. "I was astonished to see their long-term planning and research to take lithium batteries from primary batteries—the kind you just use once and throw away—and to make these things rechargeable." The visits reinforced what he already suspected: that no matter what, energy storage would one day be important. "There are certain things you just know are going to matter," he said. "Energy is going to matter to society. You're going to need energy. You're going to need to move around with energy. These kinds of fundamentals don't go away."

Elton Cairns was another among the few Western scientists still working on lithium batteries in the 1980s. He had already had a long career, leading the battery-development efforts at General Motors for its Electrovette concept car before coming to the University of California at Berkeley in 1978. Like Bruce, he watched Western companies actively cede this new industry to Japanese competitors. "Throughout the eighties, the Japanese really kept at the R & D," he said. "It took a lot of systematic work to develop a commercial lithium-ion cell: testing carbons for the negative electrode, working on procedures for making the positive electrode, finding out what would be the best electrolyte. But they did it rather quietly. Most people not inside those efforts didn't fully realize what was taking place in terms of moving toward commercialization." As for the Evereadys and Duracells, American "battery companies have never been very aggressive about doing research," he said. "They kind of got left in the dust."

In the early 1980s, Sony pursued nontoxic batteries through its subsidiary Sony-Eveready, a partnership with Union Carbide, the owners of the dry-cell giant Eveready. According to Sony's official history, the company wanted to work with Eveready on a rechargeable lithium battery, but then on December 3, 1984, Union Carbide presided over a catastrophe. A disaster at its pesticide plant in Bhopal, India, sent a forty-two-metric-ton cloud of methyl isocyanate gas wafting into town. A half million people were exposed to the poison; 2,259 people died immediately, and at least 25,000 would eventually perish as a result.

Union Carbide would be entangled in lawsuits in India for decades. In the immediate aftermath, however, the company faced another threat: a hostile takeover bid by the New Jersey chemical company GAF. To raise money, Union Carbide sold off all its consumer goods businesses, which included iconic American brands such as Glad and STP. The company sold its battery division, which through its Eveready and Energizer products held 60 percent of the American battery market and 30 percent of the world's, to Ralston Purina for $1.4 billion.

In the shadow of ten-figure corporate sales like these, the fate of Union Carbide's Sony-Eveready joint venture drew little notice. In Sony's account, Keizaburo Tozawa, head of Sony-Eveready, learned of the Union Carbide fire sale by telegram and promptly set off for the United States with a gaggle of lawyers. For a meager $12 million he secured all of Union Carbide's shares of the Sony-Eveready battery venture, on the condition that the business could continue only under a different name. Sony would call the company Sony-Energytec.

In 1987, Tozawa's group decided to focus on finally developing a mass-market rechargeable lithium battery. They had two decades of published research to work with, research by people like Stan Whittingham, Bob Huggins, Michel Armand, Elton Cairns, and John Goodenough. In the late 1980s, a few blips on the Japanese media radar had reported that Sony-Eveready was excited about the potential of a lithium-manganese-oxide battery, which would have built on a chemistry that John Goodenough had developed with a visiting South African researcher named Michael Thackeray. Those reports soon vanished, however, and were replaced by news that Tozawa's team had finally figured out how to make a lithium-ion battery safe and cheap enough for large-scale manufacturing.

The solution: a carbon anode. Using Goodenough's cobalt compound in the cathode and Sony's carbon anode, the battery operated on this simple reaction:

$$LiC_6 + CoO_2 = C_6 + LiCoO_2$$

A compound made of lithium and carbon—LiC_6—reacts with a compound made of cobalt and oxygen—CoO_2. Lithium ions flee the carbon-based electrode and swim across the electrolyte to the cathode. Once they arrive, they burrow into the crystalline lattice of the cobalt oxide, docking into place and forming a new compound, lithium cobalt oxide. Meanwhile, this reaction sends a steady stream of electrons out of the anode and through an external circuit; after leaving the external circuit, these electrons burrow into the cathode, finding their own place in the atomic Jenga puzzle that is an insertion compound battery material.

The high electrochemical potential between the two electrodes gave the battery a potent 3.6 volts, which is desirable because of this equation: $P = VI$. Electric power (P) equals voltage (V) times current (I). With a higher voltage, you can get more power with the same amount of current, and that's why it's often more efficient to run certain devices at higher voltages.

The major jump in voltage and the near doubling of energy capacity that Sony's rechargeable lithium battery delivered would radically shrink the power supply in all kinds of electronic gadgets. At the time, for example, cell phones used a 7-volt radio-frequency power amplifier to convert the phone's electrical signals into radio frequencies and then beam them to a base station. Nickel cadmium (NiCad) and even the new nickel-metal-hydride batteries, which came out in 1990, both had a nominal voltage of 1.2. That meant that it took six of the most advanced cells on the market, wired in series, to reach the 7 volts necessary to run the cell phone's power amplifier. Along comes Sony's rechargeable lithium battery: simply by virtue of its 3.6 volts, it reduced the number of battery cells necessary to power a cell phone from six to two—a dramatic reduction in the amount of physical stuff that had to be built into each phone.

Sony's new battery would last longer than any that had come before too, because the chemical reaction it relies on is extraordinarily reversible. Change the direction the electrons are flowing, and the whole thing happens in reverse. Both electrodes return to their original, untouched state. This can happen again and again with only minimal collateral damage, which means the battery can be charged and discharged hundreds of times without losing enough capacity to render it useless. Moreover, lithium ion didn't suffer from the "memory effect," a tendency of NiCad and nickel-metal-hydride batteries to permanently lose energy capacity if recharged before having been run completely dead.

The sum of these benefits made Sony's battery revolutionary. By 1988, the company was preparing its Koriyama factory to build the cells. To distinguish them from the flaming lithium-metal-based batteries that made news at Exxon in the 1970s, these would be called lithium-*ion* batteries. In February 1990, Sony made the official production announcement. After more than two decades of misfires, the rechargeable lithium battery would finally make it into the world.

Sony's announcement immediately focused attention and funding back on advanced battery research. "The commercialization of the lithium battery nailed down the argument of the detractors," who argued that lithium was too volatile ever to put in a rechargeable battery, Peter Bruce said. "There was a view from people who didn't work in the area that [rechargeable lithium batteries] had good theoretical potential, but that you couldn't make them practical. Commercialization demonstrated that you could."

The naysayers who had claimed that rechargeable lithium batteries could never be made practical had at least two good cautionary tales to back them up: Exxon, and the story of Moli Energy, a Canadian company that three years before the arrival of Sony's lithium-ion technology began selling rechargeable lithium batteries, mainly for use in cellular phones in Japan.

Moli was started primarily to find a good use for a mineral resource. The founder was Rudi Haering, a professor of physics at the University of British Columbia, who in the late 1970s heard about Exxon's titanium disulfide battery. Haering was interested in layered compounds like titanium disulfide, and he knew that his corner of Canada was, oddly enough, home to a large stash of a different but very similar layered compound, molybdenum disulfide. Molybdenum disulfide has almost exactly the same structure as titanium disulfide, but the difference is that molybdenum disulfide is a stable, hardy mineral—a compound resistant enough to rain and air that it occurs in nature. Titanium disulfide (TiS_2) is a different animal, and the contrast between the two makes Exxon's decision to kill Whittingham's TiS_2 battery sound reasonable.

"TiS_2 was a poor choice," said Jeff Dahn, a scientist at Dalhousie University in Nova Scotia who worked as a researcher at Moli in the 1980s. "You have to synthesize it under completely sealed conditions. This is extremely expensive. And as soon as you expose it to air, it stinks—it literally stinks—because the moisture in the air reacts with TiS_2 to make hydrogen sulfide. People like Stan Whittingham and whoever will tell you, 'Oh, you know, Exxon had everything figured out in the 1970s, and it was all about management screwups.' Well, not true. Their electrode material was totally unworkable." When Exxon began working in earnest on Whittingham's battery, one company set out to manufacture raw titanium disulfide in bulk. "It was like $1,000 a kilo just for TiS_2 raw material," Dahn said. "It was ridiculous. I bought a kilo of that just so I could see what it was like. Boy oh boy, open that can, and you gotta clear the room."

In 1977, with money from the mining magnate who owned rights to the region's giant molybdenum disulfide stash, Haering founded Moli Energy, its name a mashup of the elemental symbols for molybdenum and lithium. The original goal was a battery big enough to power an electric car, but as Exxon did, they had to start much smaller.

Until the mid-1980s, Moli remained a privately held company operating in research and development mode. Jeff Dahn was hired as a project leader for materials science in 1985. "When I got to Moli, it had prototype products of this lithium-molybdenum-disulfide cell that could do three hundred charge-discharge cycles or so, and they had demonstrated and shipped samples to various customers," he said. "People were showing an interest because they were far better than NiCads at the time, which were the competition."

By the spring of 1986, Moli was making four hundred of its rechargeable lithium cells each day in its R & D plant in Burnaby, British Columbia, and soon it was pitching its technology as "the first breakthrough in battery technology in almost half a century." James Stiles, a senior researcher at the company, bragged to the *Globe and Mail* that unlike unnamed previous lithium rechargeables, "our batteries don't explode." The Japanese trading house Mitsui bought the rights to sell the batteries in Japan, and the U.S. military became interested.

That same year the company held a public stock offering, which it used to finance the construction of a factory in the Vancouver suburb of Maple Ridge designed to build as many as thirty million cells a year. Problems with manufacturing equipment made putting the plant into production something of a nightmare. For a negative electrode, Moli's batteries used ultrathin sheets of metallic lithium foil; the foil needs to be stretched tight and run through machines, but "handling lithium foil is like handling a wet lasagna noodle," Dahn said. "We were the first manufacturing plant for rechargeable lithium ever. Today setting up a lithium-ion plant is trivial; you go to Japanese or Chinese equipment makers and you say, 'Give me a winder'"—the device that winds electrode foil into a roll before it's placed in the cylindrical container—"and you have a machine that works. In those days, it was the first time it was ever done." The equipment problems led to delays and the wanton burning of money.

In 1988, after plenty of technical fixes and a second round of financing, Moli's first battery, the 2.2-volt MoliCel, hit the market. Most of them went to Japan, where they were used in NEC laptops and NTT cell phones. And it turned out that while they might not explode, they burned quite well, and did so spontaneously after just a couple of months of use. In August, an NTT phone equipped with a Moli battery caught fire and injured its user. NTT recalled ten thousand phones that used Moli's battery, and Moli suspended production and began a months-long period of crisis-level safety testing.

"We were quite shocked, because we had put these things through extremely intensive safety testing," Dahn said. "And they passed them all. So in R & D we were saying, 'Holy manoley, what's happening here?' And guys like me, we were really on the hot seat.

"What we learned was that in a cell-phone application, you have your phone on standby most of the time. It's turned on and is discharging at a low rate. It will take four to five days for your phone to discharge itself, and then you'd have to charge it, and that is a situation we'd never tested the cells under." Under deadline pressure, Moli's engineers never thought to subject the batteries to this slow, tedious cycle, in which a cell is slowly drained for five days and then recharged in ten hours, and then the whole thing is repeated hundreds of times. "What happens under such a situation is that the lithium gets *extremely* high surface area," which can cause it to react violently with the electrolyte should something go wrong. Dahn emphasizes that fewer than twenty of two million Moli batteries malfunctioned—a failure rate that is "very hard to detect in testing at any level."

In October, Moli laid off 56 of its 192 employees, which was a dramatic enough move that four days later the Toronto Stock Exchange stopped trading in the company. Before long Moli was in bankruptcy. Mitsui stepped in and bought the company, and today it still exists in the form of E-One Moli Energy Canada.

It does not, however, still build rechargeable batteries using metallic lithium.

Back in the late 1980s, those who used the downfall of Moli to argue that rechargeable lithium batteries could never work didn't see the lithium-*ion* battery—which contains no metallic lithium, only benign charged lithium ions—coming. For Dahn, the lesson of Moli is that "lithium metal is completely out of control, because you have no control over what the user is gonna do."

Despite their exorbitant price, ungainly dimensions, and limited coverage, cell phones established a solid American foothold in the second half of the 1980s. Motorola sold $180 million worth by the end of 1984, the first year the devices were on the market. The number of cities with cell-phone providers grew from two in 1984 to eighty-two in 1985. By 1990, every American market had at least one cellular provider, and five million Americans subscribed to a cell phone service.

As cell phones and other gadgets proliferated, battery power once again registered in the public consciousness. "The competition is becoming fierce," a Sanyo battery-plant manager told the Associated Press. "Everybody is demanding products with longer life and less weight." In 1992 *The Economist* declared that it was about time the battery business caught up with the rest of the consumer-electronics world. "While electronics manufacturers produced ever smaller and cleverer machines, the sleepy battery business barely changed. So light have most portable devices become that the battery now accounts for a quarter of the weight, compared with a tenth a decade ago. Now, with advances in microchip technology making even tinier electronic products possible, battery makers are scrambling to come up with the lighter, more powerful and longer-lasting batteries needed to turn such gizmos into mass-market items."

In 1990, the nickel-metal-hydride battery had arrived as a less toxic, higher-energy alternative to NiCad, and it took off quickly. Almost immediately, however, it was thoroughly one-upped by Sony's lithium-ion battery. Beginning in 1992, Sony offered its new lithium-ion battery as a $60 optional power pack for the Handycam CCD-TR1 8-mm camcorder. It was 30 percent smaller and 35 percent lighter than a NiCad battery that contained the same amount of energy. It stored 90 watt-hours of energy per kilogram—triple the capacity of lead-acid batteries, nearly double that of nickel cadmium, and a good 10 to 20 percent better than nickel metal hydride. Demand for Sony's lithium-ion battery grew quickly. By March 1993, the company had shipped some three million of them, and by the following year, that number became fifteen million.

Competitors quickly entered the fray, almost all of them Japanese. Within a few years, Sanyo, Matsushita, Mitsui, Yuasa Battery, a company called A&T Battery (which was owned by Toshiba and the chemical company Asahi), and the newly Japanese-owned E-One Moli Energy were all chasing Sony in the lithium-ion race.

Meanwhile, the cell phone was working its way into the mainstream. "As recently as five years ago the cell phone was still seen as about as essential as a second Porsche," *The Economist* wrote that October. "No longer." In 1993, there were thirteen million cell phones in the United States, and they were getting cheaper by some 25 percent a year. The wireless revolution did have its birth pangs. In 1992, when the 33.5-million-circulation *USA Weekend* magazine polled its readers on their most pressing health concerns, electromagnetic fields came in at number one. January 1993, a guest on *Larry King Live* blamed his wife's brain cancer on cell-phone use; Motorola's stocks plummeted (and then quickly rebounded). As wireless carriers began building more and more cell-phone towers across the country, not-in-my-backyard protests became widespread. Even so, nothing could stop the advance of the cell phone.

Released in 1994, Motorola's MicroTAC Elite was the first mass-market mobile phone to use a lithium-ion battery—two of them, because the phone still used a 7-volt RF power amplifier. A press release trumpeting the arrival of the "world's lightest cellular telephone" used a battery analogy to put the phone's low weight in context: the entire 3.9-ounce phone was lighter than a single D-cell battery. It got forty-five minutes of talk time, six hours of standby, and it also happened to be the first phone to come with voice mail.

That a gadget revolution was imminent should have been obvious to anyone paying attention to the plummeting cost of computer chips, the growth of available wireless spectrum, and the arrival of battery technology that could at last make the pocketable sci-fi phone a reality. In 1994, *Electronic Engineering Times* identified lithium-ion batteries as the "'enabling' technology" that could "ensure the continued rapid growth of the portable electronics industry. These revolutionary batteries, which are now under development and are just beginning to emerge, will certainly be a critical factor in the proliferation of portable electronics."

In Japan, manufacturers were scrambling to meet demand. That year, Sony began building a second lithium-ion factory. NEC started expanding, and newcomers such as Mitsubishi Cable entered the lithium-ion business. In time, this early boom would result in total Japanese dominance of the lithium-ion industry. "As they did with flat liquid-crystal-display screens," wrote a *Nikkei Weekly* correspondent, "Japanese companies are securing a near-lock on producing an increasingly key electronics component—this time rechargeable lithium-ion batteries."

John Goodenough never met any of the Sony scientists who commercialized the lithium-cobalt-oxide battery. He had no idea that Sony's 1990 announcement of impending full-scale commercialization was coming. He said he has never received royalties for inventing the primary active material in the battery that now powers nearly every one of the billions of mobile phones and laptops and iPods and digital cameras in the world.

He holds no grudge against Sony, however. "I congratulate the Sony people for having the idea and doing it," he told me. "They deserve credit." He reserves his ire for England's Atomic Energy Research Establishment. He said that when he couldn't find a company interested in licensing his lithium-cobalt-oxide cathode, he turned to the AERE laboratory in Harwell, near Oxford; he was working with scientists there to obtain battery research funding from the European Economic Community. It didn't work out so well. "The lawyers at AERE Harwell swiped everything," he told me.

"I had an agreement with the people there by word of mouth that sure, you'll patent it, I'll let you recoup all the costs of patenting, and then we'll split things after that," he said. "I thought that was a reasonable offer. But the last day when we went to sign, they said, 'The lawyers will not do anything unless you sign your rights away.' I didn't know that it would be used for cell telephones, and there wasn't any battery company in Europe or England or America that wanted it at that time, so I signed my rights away." Here he laughed his bellowing laugh. Until the patent on lithium cobalt oxide expired in 2002, every lithium-ion battery manufacturer had to pay Harwell licensing fees. "They made billions," Goodenough said. "And they wouldn't even give a little bit of money to my college."

The rapid and eventually ridiculous shrinkage of the mobile phone began in earnest in 1996, with the arrival of the Motorola StarTAC. Its release was an inflection point in the history of wireless communications, the beginning of the age of the ubiquitous tiny, fashionable cell phone. Weighing only 3.1 ounces, StarTAC was the first clamshell, or "flip" phone. Unveiled at the glitzy annual Consumer Electronics Show in Las Vegas, it was pitched as the first in a new product category: the "wearable" cell phone. "Because StarTAC is so attractive and discreet," read a press release for the European version of the StarTAC, "Motorola believes that many users will WEAR rather than carry it—on a belt, in a top pocket or even pendant-style on a necklace cord."

It's unclear how many people wore the StarTAC pendant-style in the late 1990s, but the phone, which sold for between $1,000 and $2,000, did indeed become a fashion item. "A big cellular phone used to be a sign of power," read *Forbes*. "Now a small one is." *Fortune* included the StarTAC in a list of "twenty indispensable luxuries for those who travel enough to call airports home," alongside a $350 Brioni robe and a $375 Arte & Curio leather golf travel set. StarTAC was even singled out in an *Entertainment Weekly* article on the murder of Tupac Shakur as a choice gangsta fashion item: "Living the life for many high-profile rappers actually means wearing the best Versace and Moschino, downing Cristal, checking the Skypagers and dialing up their tiny $1,000 StarTAC cell phones." Unlike many fashion items, however, the StarTAC would have a lasting influence. Nine years after the phone's release, *PCWorld* ranked the StarTAC as the sixth greatest gadget of the past fifty years. (Number one was the Walkman.)

StarTAC's tiny size was made possible both by lithium-ion batteries and an accompanying advance—the affordable 3-volt RF power amplifier. "Now you could get away with a single lithium-ion cell to run the phone," said Jason Howard, who started at Motorola as a battery researcher in 1993. "You get out of all the problems associated with putting cells in series to get to the required voltage." Together, the high-voltage, high-energy ultralight lithium-ion battery and the lower-voltage

RF power amplifier meant that cell phones had gone from running on six batteries to one in just a few years.

By 1997, there were more than 120 million cell-phone subscribers in the world. In a year, that number would more than double. According to an industry publication, new subscribers were signing up at a rate of two each second. The lithium-ion market, meanwhile, had been growing at a clip of 200 to 300 percent per year.

By 1999, the cell phone had begun its transformation into the smart phone, the handheld everything-device that we've grown accustomed to today. That year the Japanese phone giant NTT DOCOMO came out with an Internet-capable phone called i-mode. The merging of the cell phone and the Web-connected Internet led *The Economist* that October to trumpet the "conquest of location." "Until recently, one of the most bothersome aspects of the information age was that people risked becoming powerless whenever they left their home or office," the article read. "Now mobile phones are putting communications in their pockets."

A piece in *Time* the following year was breathlessly enthusiastic about wireless. After noting that Qualcomm had seen its stock rise more than 3,000 percent in just over a year, the article explained that "we're on the brink of a major technosocial upheaval that's right up there with the steam engine, car, and computer. It promises the ultimate technological breakthrough for the information age. Virtually all information will be available to you at all times, whether you're taking a day off from work, visiting the in-laws or traveling to Fiji. With the importance of physical location diminished, even irrelevant, you'll be able to answer an e-mail from your boss, shift your 401(k) or sing your child a video-and-sound lullaby wherever you are."

It didn't take long for people to realize that always being connected had drawbacks as well. "Perhaps the more worrisome outgrowth of the wireless Web is the never-ending workday," warned *BusinessWeek* in 2000. "Cell phones and the Net are already stretching work at both ends. The marriage of the two technologies puts the trend on steroids . . . And while it's easy to say workers should simply turn off their machines, those that do may well find themselves competing for promotions and bonuses with others who don't."

Not that concerns like these had any effect. By 2002, cell phones, and with them, lithium ion, were everywhere. Some 95 percent of the cell phones on the market that year, regardless of cost, used lithium-ion batteries. The once exotic technology had established itself in less than a decade as the standard power source for consumer electronics. In 2002, *The Economist* anointed lithium-ion batteries the "foot-soldiers of the digital revolution," comparing the cell phones of the day to the Mobira Senator, the 21.6-pound cinder block of a car phone Nokia unveiled in 1982. "Today, a typical mobile phone is a hundredth of this (i.e., 100 grams or less) and can be tucked discreetly into a shirt pocket," read the story, headlined "Hooked on Lithium." "This 99% weight reduction has been achieved largely through advances in battery technology. Above all, it is down to one particular breakthrough: the advent of the lithium-ion rechargeable battery."

4

REVIVING THE ELECTRIC CAR

Inevitably, the lithium-ion battery leaped into the automotive realm. Martin Eberhard, 1990s e-book entrepreneur turned 2000s electric-car impresario, was an essential early actor in this technological transfer. A Silicon Valley electrical engineer, Eberhard first encountered lithium ion in 1996, when he and his partner Marc Tarpenning started the e-reader company NuvoMedia. They wanted to put it in their first product, but in those days lithium-ion batteries were still delicate and not fully understood, and the battery companies were selective about whom they'd sell to. Lithium ion required more care and handling than the existing standard, nickel metal hydride; it needed voltage balancing, and if that wasn't done properly, the batteries could become dangerous. By the time NuvoMedia was building their second-generation reader, they found a willing supplier for lithium-ion cells, as long as they sent some engineers to lithium-ion charging school and submitted their control-circuit designs for approval. They acceded to the demands, and they were glad they did. Eberhard said he found lithium ion to be a major improvement over everything that had come before.

In 2000, Eberhard and Tarpenning sold NuvoMedia for $187 million. Flush with cash, they started thinking about their next business venture. "I was pushing for electric cars for a lot of reasons—political, ecological, and so on," Eberhard said. "Lithium-ion batteries were at the top of my mind, because in my rough calculations you could actually fit enough batteries into a car to make a meaningful car." Meaningful meant more than two hundred miles of range on a charge. It meant a car that would bury the reputation of electric cars as sub-golf-cart transport suitable only for gated neighborhoods.

Eberhard and Tarpenning sketched out their idea for a lithium-ion-powered car and convinced themselves that it was feasible. Coincidentally, Eberhard had just put some of his NuvoMedia earnings to work rescuing the boutique Los Angeles car builder AC Propulsion, creators of a lead-acid-powered electric sports car called the tzero. Soon Eberhard was asking them about lithium ion—ever think of putting that in a car? "At first they didn't want to talk to me about it," he said.

Probably because they had been thinking the same thing. Before long, AC Propulsion came around and started comparing notes with Eberhard. "They had this idea of basically Krazy Gluing the cells together and attaching them with connectors at the top and the bottom," Eberhard said. What they needed, in addition to the advice of an electrical engineer, was money, and both of those things Eberhard could provide. Together, Eberhard and AC Propulsion built a lithium-ion-powered tzero, and it was just as cool as Eberhard had hoped—a one-off, sure, but still a purely electric sports car that darted from zero to sixty in 3.6 seconds and ran up to three hundred miles on a charge. Electric drive and high performance were no longer mutually exclusive.

Eberhard wanted a lithium-ion tzero for himself, even if he had to start a car company to get it. AC Propulsion, however, had no interest in becoming a full-fledged carmaker. Eventually they even begged off of building Eberhard his own tzero, saying they didn't have the resources. "I suppose if they had actually done that," Eberhard said, "I wouldn't have started Tesla Motors."

On July 1, 2003, Eberhard and Tarpenning incorporated Tesla Motors with the goal of building the hottest electric cars the world had ever seen. Eberhard was unfazed by the recent bursting of the 1990s electric-car mini-bubble, in which GM, Toyota, and a handful of other car companies built small fleets of electric cars, only to drop the programs when they succeeded in killing the pollution laws that inspired them. Not only was he going to use technology far superior to anything tried before, but now that everyone else had gotten out of the electric-car business, he had the field to himself. He began courting investors. "Part of my convincing them was that this car"—the lithium-ion-powered tzero—"shows that you can make an electric car that's fun to drive," he said. "And we're in this unique position where all the people who would otherwise be competitors have bailed out of the market."

The idea behind Tesla was to compete on performance rather than price. At a higher price, they could afford a $20,000-plus battery pack, and they could use an advanced electric motor. By starting at the top of the market, they could show what electric cars could do—what they could, with twenty-first-century technology, really be.

Immediately after incorporation, Eberhard and Tarpenning rented a two-suite office in Menlo Park with room for five or six people. They had no garage, but that was okay, because they weren't anywhere near ready to build a car. It was all design work on computers, refining a business plan, and searching the world for partners, hunting for those rare suppliers who, although accustomed to dealing with the likes of GM and Ford, would be willing to meet with a couple of guys who had a PowerPoint presentation and a dream.

They got their first big break at the end of the year. "Marc and I went down to the LA Auto Show and basically forced ourselves upon the Lotus people and made them listen to our story," he said. "And they liked it. We had a handshake deal that they'd work with us." Once they knew they were working with a Lotus platform, they could begin computer modeling of the car they hoped to build. They could start finding concrete answers to essential questions. How many batteries will fit in a roadster based on a Lotus platform? What kind of range will that provide?

They soon arrived at the design for their battery pack, a conglomeration of 6,831 laptop cells. It would take an incredible amount of work to turn this idea into a production-ready battery pack. They had to carefully measure the efficiency of each cell, because inefficiency means heat, and heat can mean disaster. What was the best way to get rid of the heat from the cells? They tried liquid cooling, they tried air cooling, they fought and gave each other ulcers about the decision. They conducted computer simulations. They watched cells in laboratory conditions using thermal-imaging cameras. Finally, they settled on a liquid-cooling system.

Then a potentially show-stopping problem arose. "About a year into our design, some of our engineers went to a battery conference where a couple of people had presented papers talking about the possibility of thermal runaway in a lithium-ion cell," Eberhard said. "And that was a surprise to us, actually. If you read the data sheets from those days from the cell manufacturers, they talk about the cells' being safe when you penetrate them and overcharge them and crush them. Every way you read it, you think there's nothing that could go wrong with these cells. And these papers that were presented suggested otherwise." Before long, word of the occasional laptop fire started spreading across the Internet, and Eberhard and his team decided to conduct some experiments of their own.

For the first test, Tesla engineers went to the parking lot and laid down a single laptop cell (called an 18650 cell because of its dimensions), wrapped it with nichrome wire (the highly resistant wire used in toasters), and cooked it. "It went off like a Roman candle," Eberhard said. Next they dug a hole in Eberhard's yard and buried a group of battery cells that were wired together and outfitted with thermal probes, voltage probes, and a video camera. Then they set one of the cells on fire by overheating it. "It was quite dramatic," Eberhard said. The cell eventually became an inferno and set its neighbors aflame too, starting a full chain reaction. "That was very scary to us."

Eberhard announced to the board that they would stop production of the car until they had solved the battery safety issue, and before long, they did. "We learned that if you design the battery system right, you can design it so that if one cell burns it doesn't catch anything else on fire," he said. "You can prove that through testing and through modeling." Still, the battery safety pause delayed the company by months.

Fortunately, they had enough funding to endure such a delay. In February 2004, they had persuaded Elon Musk, a cofounder of PayPal who in 2002 had founded the space-rocket company SpaceX, to join them as a major investor. As a condition of his investment, he would be appointed chairman of the company. Musk presided over several subsequent rounds of funding, and he convinced fancy friends to invest, including Sergey Brin and Larry Page of Google and eBay's Jeff Skoll.

The delay still hurt. The Tesla team had been obsessed with secrecy, going so far as to code-name the car that they would eventually christen "Roadster," the generic name for any convertible two-seat sports car. (Instead, the Roadster was "Dark Star.") Yet Tesla revealed its car to the public before anyone would have liked because they felt it was essential to prove that the company was serious. They had to "separate ourselves from the AC Propulsions of the world," Eberhard said, in order to attract better engineers and to get big-time automotive suppliers to take Tesla seriously. "Just imagine, picture yourself back in 2005 or so. You call up Siemens: 'We'd like to buy some air bags from you. Custom design one for us, by the way.' They'd laugh in your face, a bunch of fruits and nuts in California. There's just no way that would happen," he said. "The idea of becoming public was to show that what we had really was going to go into production."

The launch happened in a hangar at the Santa Monica airport on the clear California evening of July 19, 2006. Michael Eisner and Arnold Schwarzenegger showed up, as did the core of the West Coast electric-car intelligentsia—the investors, the documentarians. Eberhard was effervescent.

Low-slung and sleek, essentially a Lotus tastefully electrified, the Tesla Roadster was by far the sexiest electric car anyone had ever seen. For hours, attendees were given rides around the airport tarmac. A few big shots were even allowed to drive it. The car bolted from zero to sixty in four seconds, putting it in Porsche and Ferrari territory. It did so in silence, powered by the electrons coursing from its massive battery to its custom AC induction motor.

The reviews came in quickly, and they were exuberant. Even the editorial board of *The New York Times* got caught up in the excitement. In an editorial with the headline "Go Speed Racer!" the board wrote that "Tesla Motors, a Silicon Valley start-up, has developed a two-seater that goes from zero to 60 miles an hour in four seconds, leaving the days of electric cars as glorified golf carts in the dust." Later that year, *Time* would name the Tesla Roadster one of the best inventions of 2006.

In Detroit, Bob Lutz followed the enthusiastic press coverage of Tesla with growing frustration. This was not a good year for the reputation of General Motors, where Lutz, a towering seventy-four-year-old, was head of global product planning. The company was drowning in cost. The health-care and pension deals it had struck with the United Auto Workers over the decades meant that the company was spending some $6 billion a year on benefits. At the same time, its concentration on big SUVs and trucks was backfiring as drivers began to move away from giant gas-guzzlers. GM was operating according to a business formula that left no room for a sudden spike in oil prices or a shift in driver preferences. As its products became less profitable, its obligations to its employees became more expensive. In 2005, CEO Rick Wagoner was in the middle of a major restructuring of the company, cutting costs and renegotiating health care with the union. But it wasn't enough. In the first quarter of 2005, the company lost $1.1 billion. That number would soon be dwarfed by the almost incomprehensible losses GM would register in the coming years.

In January 2006, *Who Killed the Electric Car?* premiered at Sundance, and by the summer it was showing in theaters and bound for sleeper-hit status. The documentary helped harden the perception of General Motors as chronically insincere in its alternative-fuel efforts.

Then there was the Prius. When Toyota began selling its famous hybrid in the United States in 2001, Lutz and others at GM dismissed it as a marketing stunt. An awkward little econo-box that uses an expensive nickel-metal-hydride battery to squeeze out a few more miles to the gallon? Please. This was not a real car. This was not a threat. General Motors management persisted in believing that the Prius was not a threat until it was. The Prius gave Toyota an extraordinary image boost. By the time the Tesla and *Who Killed the Electric Car?* came out, Toyota was on its way to surpassing General Motors and becoming the world's largest automaker.

Lutz had actually been fighting for a counterpunch of an electric-car project for some time. "I started proposing that we do something all electric with lithium-ion batteries about the time that Toyota was really starting to reap the PR benefit from the Prius. That would've been around 2004," Lutz told me. "My original concept was more like the Nissan Leaf. In other words, a huge 25-kilowatt-hour lithium-ion battery to provide a range of about a hundred miles. And that was roundly rejected because of everybody remembering the misery of the EV1 and the bad press we got."

Lutz also encountered this argument: We already have hydrogen fuel-cell vehicles! GM was indeed spending vast amounts of money on hydrogen, the alternative-energy red herring of the early 2000s. But there was a growing realization that the infrastructure required to replace gasoline with another liquid fuel—one that, incidentally, was most often derived from natural gas—would be staggeringly expensive and take decades to build. "I said, 'Well, to be honest, nobody really takes that as seriously as they should because it is too easy to say, "GM is working on a technology that will be very difficult to generalize because of the absence of a hydrogen-refueling infrastructure,"'" Lutz said.

Nothing less than an electrically driven car would appease these critics. Lutz's argument was, "Whether we like it or not, there are the environmentally conscious people, the government people in California, the car people—everybody seems to believe that the electric vehicle is the answer. And with the advent of lithium-ion batteries, I think that we're getting to the point where it is feasible."

His idea was killed again on technical grounds. "I had some battery experts explain to me that lithium ion was not suitable for automotive use because it is an energy battery, not a power battery. It's okay for powering laptops and devices, but you can't make a power battery out of lithium ion, et cetera et cetera. So I kind of grudgingly let it go to sleep for a while."

Then he started hearing rumors about the Tesla Roadster.

"I had an outburst at one of the big automotive strategy committee meetings. I said, 'How is it that everybody at GM convinces me that this can't be done, and we're supposedly the world's most competent car company, and I have every expert explaining to me that lithium batteries won't work. And here is this outfit in California that nobody has ever heard of, and they are gonna put a car on the market with lithium-ion batteries, it's gonna accelerate from zero to sixty in four-point-something seconds, gonna have a top speed of 140 and a two-hundred-mile range?'" The response: Don't believe all that hype, Bob. "I said, 'All right, discount everything by 20 percent and it is still sensational.' And of course, we didn't know about the $100,000 price tag back then, or the fact that the production would be so limited. But to me, it was the signal that, hey, there is a way to get there with lithium-ion batteries."

Lutz, a four-decade veteran of the automotive business, had already had a remarkable career. It began in the golden years of the early 1960s. Would it end with the greatest car company the world had ever known disgraced and teetering on the brink of bankruptcy? By 2006, Lutz had become a symbol of Detroit, an astonishingly candid, gruff-voiced cigar smoker and whiskey drinker, an old-school charmer, an icon from the Good Old Days. An ex-marine born to a Swiss banker in 1932, he famously flew fighter jets on the weekend and drove a Corvette to work every day.

Educated at the University of California at Berkeley before serving as an active-duty marine from 1954 to 1959, Robert A. Lutz got his start in the car business at General Motors of Europe in 1963. In 1971, he decamped for a three-year stint at BMW, where he was in charge of global sales and marketing. Next came a twelve-year tenure at Ford. By then he was becoming known as one of the most colorful men in Detroit, earning a place in David Halberstam's epic story of the battle between Ford and Nissan, *The Reckoning*. In 1986, Lee Iacocca recruited Lutz to help rescue Chrysler. In his twelve years there, his crowning and most personality-appropriate achievement was the production of the Dodge Viper, a V-10-powered hell demon of a two-seat sports car.

Next he spent four years at Exide Technologies, the lead-acid battery manufacturer descended from the Electric Storage Battery Company, which Thomas Edison spent the first decade of the twentieth century racing against. Lutz's time at Exide gave him a strong appreciation for the battery. "Exide was never a large producer of advanced batteries, but we did have a subsidiary in Germany that did very exotic batteries for space applications," he said. "For instance, batteries that would power tiny stepper motors that would make minute adjustments on space telescopes or space lasers. And these were batteries that had to hold their charge for ten to fifteen years, because it's tough to get up into space to swap batteries. We had all kinds of advanced primary batteries for munitions for the NATO military. I found that fascinating."

In 2002, he returned to GM. Four years later, the company was a disaster, but the downfall had been a long time coming. GM was crippled by decades of accumulated health-care and pension obligations and hopelessly dependent on big SUVs and trucks, which were inordinately profitable but reliable sellers only in an era of reliably dirt-cheap oil. Any shift in consumer preference would expose GM as dangerously dependent on the likes of its Hummer. That, of course, is exactly what came to pass.

Fixing General Motors would require renovating the culture and corporate organization, taking a thresher to the company's cost structure, and rebuilding the company's ruined reputation among car buyers. As head of global product planning, fixing the company's reputation with good cars and advanced technology was Lutz's job.

In early 2006, he and a handful of other high-level executives decided to begin work on a project unimaginatively but tellingly called iCar. Lutz enlisted Jon Lauckner, vice president for global product development, to work up the new car. Lauckner is a tall, vaguely haughty engineering savant who started his career as an employee-in-training with Buick in 1979 and worked his way up to vice president of global program management for GM. When he's on his best behavior, he gives the impression of actively restraining himself from vocalizing the constant stream of withering retorts that are running through his brain.

The iCar was a vessel for every engineer's favorite alternative propulsion system. Ethanol fans said it should be an ethanol car. Hydrogen believers argued that it needed a fuel cell. Despite the EV1 debacle, GM still had electric-car lovers among its ranks, but a purely electric car seemed too limited. Lauckner is credited with the idea that united these warring factions: building an electric drive system with a dedicated spot for backup power.

"I sat down with Jon Lauckner," Lutz said, "and he convinced me, you do something with a huge battery, it's gonna cost a fortune, it's gonna be extremely heavy, and you'll have a limited range. The far better proposition would be, let's go for the eighty/twenty solution"—a battery pack big enough to cover 80 percent of the people, plus gas power for backup. "I said, 'Boy, that sounds real good.' Jon did some fast calculations and figured that somebody who drives sixty miles a day would probably be looking at a real-world 150 miles per gallon. Well, that got me very excited. So we talked to design and started doing some sketches of the original prototype Volt, and we talked to some battery suppliers, just enough to know what an automotive cell would look like."

The decision was to pair lithium-ion batteries with a small gasoline "range-extending" engine that would generate electricity once the battery was depleted. The batteries would store enough energy to travel forty miles before the backup gas engine was activated. After examining the market research, the Volt group realized that forty miles a day would cover 78 percent of American drivers—those people could theoretically never use gasoline. Everyone else would get gas mileage that was simply unheard of.

Lutz took Lauckner's idea to the boss. "I talked to [CEO] Rick Wagoner a few times and said, 'Rick, look, we'd like to do this concept car for the show, just a concept car. It's not a pure electric, but it would introduce the concept of lithium ion,'" Lutz said. "I would say there was grudging acceptance of the idea. And one of the reasons there was acceptance was that the board of directors, which contained some people who were very much into technology, like Kent Kresa, the former CEO of Northrop [Grumman], were actively pushing: 'When are we going to demonstrate technological superiority to Toyota? Why are we sitting back and taking this rather than making some sensational move?' So I think that pressure from the board and my constant wheedling finally got, 'Okay, all right, let's do a concept car.'"

Bob Boniface, director of GM's Advanced Design studio, had just finished his design for the revived Camaro and was enjoying the post–Detroit Auto Show lull, an annual breather before the design team begins working in earnest on the next year's show cars. "In early '06, word came down that next year Lutz wanted to do a 'technology-based vehicle,'" Boniface said. "So I ask, 'Does he want this, does he want

that?' And they say, 'I don't know, just a technology-based vehicle.'"
Boniface rolled his eyes and shrugged his shoulders, palms in the air.
We were talking the evening before the beginning of the 2010 New
York International Auto Show, standing next to a finished Volt prototype. "So we started working up all kinds of crazy stuff—maglev cars,
cars without wheels. And then I finally get some clarity: 'No, you don't
understand: Lutz wants an *electric* car.'"

Even before the Volt team had its first official employee, rumors
about the car started leaking. Chelsea Sexton, one of the stars of *Who
Killed the Electric Car?*, began hearing rumors of a new GM electric in
March 2006. "One of my friends [inside GM] said, 'We're working on
something. I can't tell you what it is, but it would make you really happy,'"
she said. "It was very clear that it had a plug on it, but he wouldn't tell
me any more than that." As it happens, that was the month that GM's
Automotive Strategy Board approved the Volt.

In April, Tony Posawatz, a short, stout engineer who looks like he'd
make a good high school wrestling coach, was hired to lead the Volt
program. He was employee number one. Posawatz faced a dilemma
that other GM engineers had faced before him: Do I sign up to lead an
electric-car project, knowing that historically GM electric-car projects
are doomed and possible career killers? He hesitated. He doesn't do
concept cars, he protested. Lauckner and Lutz assured him that while,
yes, the Volt was a concept car, it wasn't a *concept* concept. "I never
looked at it as a concept that we weren't going to execute," Lutz said.
"One of the reasons I didn't is that I knew, I knew in my gut, that the
concept of a lithium-ion-powered vehicle with a forty-mile electric range
and then another, say, almost three hundred miles beyond that, was
going to be a very, very compelling proposition."

Soon Boniface left Advanced Design to lead the Volt design team.
Three GM design studios—Boniface's studio in Warren, Michigan, another in Hollywood, and another in Coventry, England—began working
up concepts. Boniface's studio submitted five themes, the English studio
did two, and the Hollywood studio did one. In time, a theme from Boniface's team was chosen as the show car.

The Volt project began attracting veterans from the EV1 days, engineers like Andrew Farah and Jon Bereisa. Meanwhile, in the outside

world the whispers about GM's secret electrified car were getting louder. Sexton recalls standing with crew members on a Minneapolis street corner at the end of the film tour for *Who Killed the Electric Car?* Someone's cell phone rang. It was a reporter. "They said, 'Do you want to comment on the fact that GM is about to unveil a plug-in car at the end of the year?' And we thought: *Interesting*." The information aligned with everything Sexton and her colleagues had been hearing.

In November, Rick Wagoner teased the Volt during a speech at the Los Angeles Auto Show, emphasizing GM's commitment to alternative fuels and hinting that something big was on the way. That same month, as production engineering on the car began, GM held an advance press briefing to get journalists accustomed to the idea of the Volt—that it wasn't exactly an electric car, but it also wasn't a normal hybrid, and it was certainly not a joke.

GM was set to unveil the Volt concept at the 2007 North American International Auto Show in Detroit, but the company remained dangerously noncommittal about the car. Days before the show, GM executives were still debating whether it made sense to put the Volt into production. Lutz knew that it would be a public-relations catastrophe to show the car and then never build it, so he gave his colleagues an ultimatum: If we're not going to build it, we can't show it. We have to decide right now. Ultimately, of course, the show went on. At the unveiling, the official status was that GM was "actively studying production." "We all decided that it was just one of these programs where, this is where GM can finally demonstrate to the world that when it comes to advanced propulsion technology, nobody else in the world can lay a glove on us," Lutz said. "It became an important act of corporate will and demonstration of the corporate mastery of technology. And I think in this area everybody else is three years behind us."

The Volt reveal was over the top even by North American International Auto Show standards. It began with a video projection on a large rounded screen behind a stage. Images of cuneiform tablets gave way to Gutenberg's printing press before flashing forward to a twentieth-century newspaper press and then cutting all the way back to the Lascaux cave paintings. "Can you imagine modern life without the invention of the printing press?" a deep male commercial-ready voice-over intoned. "The photograph? The automobile?" Here the video raced through the history of the automobile as told by GM, pausing on an image of the EV1. "And yet, the history of innovation is largely one of evolutionary improvements." A space shuttle blasted off. A space station orbited Earth. "When one change inspires countless others, setting in motion events that will forever alter history." Neil Armstrong and Buzz Aldrin hopped around on the moon. Benjamin Franklin's kite hovered. Thomas Edison's visage covered the screen for a few seconds, providing an unsubtle hint about where this was going. "It only takes a single idea to spark the flame that lets us see all the challenges in new eyes and a better world in a flash of light."

Rick Wagoner stepped onto the stage. "About six weeks ago at the Los Angeles Auto Show, I delivered a speech on a very important topic for all of us: the inevitability and the promise of energy diversity," he began. "I highlighted some serious concerns about sustainable growth, the environment, and energy availability and supply, issues that have come to be called energy security. I made the point that it is highly unlikely that oil alone will supply all the world's rapidly growing automotive energy requirements."

The language Wagoner used to describe the implications of this energy crunch suggested that GM had decided to deny that they had ever turned away from electric cars. "At GM this means we'll continue to improve the efficiency of the internal combustion engine as we have for decades," he said. "But it also means that we'll dramatically intensify our efforts to displace traditional petroleum-based fuels by building a lot more vehicles that run on alternatives such as E85 ethanol, and very importantly by significantly expanding and accelerating our commitment to the development of electrically driven vehicles."

At the Los Angeles show, GM had unveiled a plug-in hybrid version of the Saturn Vue, due to arrive whenever the lithium-ion batteries were ready. Wagoner went on, "Now we're making another significant and important step in our commitment to an electrically driven future. Today I'm pleased to announce that GM has begun production work on

E-Flex, a family of electrically driven propulsion systems"—a notable advance, Wagoner said, because of the wide variety of fuels that could be used. "And now it's my pleasure to introduce the latest evidence of that commitment. A dynamic concept vehicle based on E-Flex technology, a sleek design that's already generating a lot of electricity in its own right."

The room began to rumble as the screen split in two. Blue lightning bolts and sparks and zaps and flashes overwhelmed the stage before giving way to a sound track of gentle electronica. A disembodied female voice announced: "Ladies and gentlemen, the 2007 Chevrolet Volt Concept."

The car looked great—low and brawny, with a wide stance, a glimmering silver paint job, and a long hood that in the visual language of gas-driven cars telegraphed power. Bob Lutz was sitting shotgun. He stepped out of the car and shook Wagoner's hand, and the two posed before the Volt for photos. Then, standing beside his baby, Lutz began speaking into his lapel mike. "Well, here it is," he said. "The Chevrolet Volt, an electrically driven car from General Motors. I am shocked, truly shocked. A GM electric vehicle is an inconvenient truth." The audience greeted the reference to Al Gore's recently released movie with awkward laughter.

Lutz described the drivetrain and the specifications. Forty miles on lithium-ion batteries alone. If you're one of the 78 percent of Americans who live within twenty miles of work, and you charge the car every night, "You will never need to buy gasoline during the entire life of the vehicle," he said, and that line drew enthusiastic applause. "And you would save five hundred gallons of gasoline and eliminate 4.4 metric tons of carbon dioxide a year from the tailpipe." If you drive sixty miles a day, you'll get the equivalent of 150 mpg, he said. Use E85 and you'll get more than five hundred miles per equivalent petroleum gallon. "I'm honestly as excited and as passionate about this program as I have ever been about anything that I've done in my forty years in this business." Lutz finished his remarks and welcomed to the stage Jon Lauckner and Tony Posawatz—"huge driving forces behind this"—and, as is ritual after these press conferences, the reporters and photographers in the audience swarmed, crowding around the GM principals to get a quote or a shot or a few minutes of on-scene footage.

The press devoured the Volt story. A GM spokesperson told the *Detroit Free Press* that the Volt reveal earned ten times the attention of a normal auto show press conference. Yet much of the coverage exhibited excitement about the car as a concept—and only as a concept. Given GM's dismal track record with electric cars and the theoretical nature of the vehicle, it was hard to take the Volt too seriously. The common refrain: Nice-looking concept car, but did you just say it runs on batteries that don't yet exist? A photo caption from *The New York Times*'s online coverage of the show put it wryly: "The Volt is truly the car of the future. It's a theoretical 150-mile-a-gallon automobile based on batteries that are not yet available."

Toyota, which couldn't abide the Volt stealing any glory from its Prius, was happy to help fuel the skepticism, and they based their argument on the batteries. In various interviews, Toyota executives said that lithium ion was too dangerous, too expensive, and worst of all, there wasn't enough readily accessible lithium in the world to support the auto market. GM executives tried to counter the cynics, with mixed success. "This program is not a public relations ploy," Jon Lauckner told the *Times*. "We are dead serious about taking this technology into high-volume production."

"You saw what the reaction was," Lutz said to me. "It was incredible. Both positive and negative. I mean, it was generally first a huge wave of positive press, and then there was the acid reflux."

5

THE BLANK SPOT AT THE HEART OF THE CAR

Skepticism about the Chevy Volt wasn't based only on suspicion of General Motors's sincerity. It was also about the fact that the history of the electric car is a parade of extraordinary failures stretching back to the late nineteenth century.

The idea that electricity could be used as a locomotive force emerged in the 1830s, when early experiments with electric motors proved that the technology could move a rail-riding vehicle. Electric motors, however, were rare at the time, and it took another three decades for them to become common. Their spread was the result of the arc-lighting boom, the rapid increase in the number of public light sources that ran on steam-powered dynamos. It's a very short leap from the dynamo to the electric motor, and as Michael Brian Schiffer puts it in his history of the early electric car, by the late 1870s, "All at once, it seems, many people on both sides of the Atlantic began to appreciate that the components needed for electric traction were nearly at hand."

In the late nineteenth century, American and European scientists and engineers were busily inventing new ways to move people around without horses, which filled the streets with manure, were difficult to care for, and sometimes simply died on the spot. The bicycling craze began in France in the 1860s, and by the 1890s it had spread to America, where the bicycle gave city dwellers a taste for "touring"—riding out of town to see the countryside. In 1859, the Belgian engineer Étienne Lenoir built the first usable internal combustion engine, and in the late 1870s the German engineer Nicolaus Otto built his four-stroke "Otto cycle" engine, which found wide use in automobiles. Karl Benz patented his seminal Otto cycle–powered three-wheeler in 1886, and by the 1890s companies such as Peugeot were also selling gas-powered cars.

But the electric trolley, not the gas-powered car, led most directly to the advent of the electric vehicle. The first complete system of electric trolleys, the work of the American inventor Frank Julian Sprague, was installed in 1887 in Richmond, Virginia. Trolleys quickly became popular in cities across the United States and Europe, but then came a backlash: in dense urban areas, particularly Manhattan, the tangle of overhead power wires from trolleys (in addition to the wires from emerging electric-lighting networks) became an eyesore and a major public safety concern. The natural alternative, of course, was to power trolleys using batteries. Power a trolley by battery, then take it off the rails and let it roam wherever the driver wants to go, and you have an electric car.

The first electric passenger cars were creaky open-air tricycles, like the 4-mph banshee that the British inventor Magnus Volk revealed in 1888. A man named A. L. Riker is given credit for creating America's first electric car, a trike that doubled the top speed of Volk's vehicle. A pack of inventors soon began chasing the electric car. At the 1893 World's Columbian Exposition, the only car on the grounds was a 20-mph electric invented by an Iowan named William Morrison.

Electricity was still basically magic in these days. The *Wabash Plain Dealer* described the unveiling of an arc-lighting system thusly: "Men fell on their knees, groans were uttered at the sight, and many were dumb with amazement." With electricity still such a novelty, it's not surprising that people were slow to consider the idea that vehicles powered by this strange force could replace the horse. Yet reporting on the Columbian Exposition, the magazine *Electrical World* made this prediction: "The day does not seem so very far distant when carriages as well as other vehicles will be moving around our streets propelled by electric motors that receive their current from concealed batteries, and therefore effect a further emancipation of the millions of animals now performing this service."

Throughout the 1890s, electric, gas-powered, and steam cars all improved. In 1895, the brothers J. Frank and Charles E. Duryea built the first American gas car of note, and the following year a young Henry Ford began dreaming of the invention that would soon make him rich. In 1894, two Philadelphia inventors, Pedro Salom and Henry G. Morris, built the 4,250-pound Electrobat, which could reach 15 mph and, if driven gingerly, run as far as one hundred miles on a charge of its sixteen hundred pounds of batteries. Salom and Morris soon founded the Electric Carriage & Wagon Company, which was associated with Philadelphia-based Electric Storage Battery Company (ESB), which owned another fledgling electric-car concern, the Electric Vehicle Company (EVC). Soon Salom and Morris would build a radically improved Electrobat 2, which was two-thirds the weight and reached 20 mph. This coalition—Salom and Morris, ESB, EVC—launched electric taxi service in a number of U.S. cities, with the largest enterprise in New York. The group also had a reputation for patent-sharking and monopolistic behavior, which is why it earned the slanderous nickname, drawn from the element found in ESB's batteries, of Lead Trust.

The bicycle magnate Albert Pope saw where things were headed and wanted in. And so on May 13, 1897, in Hartford, Connecticut, he unveiled his Columbia Electric Phaeton Mark III, an eighteen-hundred-pound vehicle made light by hollow, bicyclelike frames, and propelled up to 15 mph in a thirty-mile radius by 850 pounds of lead-acid batteries. According to Schiffer, the Mark III was "the first American automobile of any kind to be produced commercially in more than trivial numbers."

The electric cars of the day were extravagances, some costing up to $5,000 (nearly $130,000 today). But that was the passenger-car market. By 1900, electric delivery wagons, trucks, buses, ambulances, and taxis were roaming city streets across the country. Still, charging was a problem, particularly for people with AC power, as was battery maintenance, which involved manually maintaining the levels of water and acid in the electrolyte. As Charles Duryea told *Horseless Age*, "a set of batteries was worse to take care of than a hospital full of sick dogs."

Meanwhile, drivers, mostly men, were coming to enjoy the freedom of touring the countryside in gas-powered cars, never worrying about the lack of a charging station (you could get gas at the general store). And as touring became the primary reason people bought cars, the electric vehicle became less attractive. As Duryea, ever the gasoline champion, once phrased it, a horse or even a bicycle would take you farther in a day than an electric car. Gas cars were getting faster and more powerful. To stereotype, men didn't mind the noise and the fumes, and gas cars became masculine touring cars. Electrics were for women— clean, quiet, refined. Even the Electric Vehicle Company eventually started building gas cars.

Then, on June 16, 1903, Ford Motor Company was born. The Model T, the legendary everyman's car introduced in 1908, would forever change transportation. By 1910, there were nearly half a million cars registered in the United States. Most were powered by gasoline. A year later, Charles Franklin Kettering invented the electric self-starter simply by doing what the experts of the day said was impossible—overstraining an electric motor for the brief moment it took to turn over the gasoline engine, which worked fine as long as the motor could rest between exertions. When a self-starter appeared on a gas-powered Cadillac in 1912, it eliminated one of the electric vehicle's last remaining advantages.

Many in the electric-car industry realized what was happening, and they responded with an advertising blitz designed to convince the public of the electric's inherent superiority to the gas-powered car—of its reliability, silence, cheap upkeep, and at last, its decent range per charge, its ability, finally, to "tour." (That, at least, was the argument. Everyone was aware of the drawbacks of limited range and long charging times.) Infighting began in the industry. Electrical utilities complained that electric cars weren't catching on sufficiently because they were too high-end—that what was needed was a Henry Ford and a Model T for the electric market. Carmakers liked to point out that electric utilities, instead of taking the obvious step of advertising their core product by using electric vehicles in their own commercial fleets, too often relied on the gas-powered cars that were robbing them of potential profits.

Before long, electric cars were stigmatized as the aristocratic grandmother's car. World War I put all but the strongest car companies out of business. "That any firms lasted past the war is testimony to the wealthy woman's loyalty to the electric car," as Schiffer put it. The last proper, commercially available electric cars were off the market by the beginning of the 1930s. "In the end," Schiffer writes, "there was no Henry Ford of the electric car because there was no mass demand for inexpensive electrics."

As the gas car colonized the world, the electric vehicle went into deep, cryogenic hibernation. Aside from isolated oddities such as the 1959 Henney Kilowatt, a production electric car that the National Union Electric Company built with the Eureka Williams Company and Exide in a vain attempt to resurrect the technology, the electric car was nothing but a wistful longing for idealistic engineers. Fewer than fifty Henney Kilowatts were built.

Not until air pollution became an issue in the 1950s and 1960s did the major automakers even pretend to show an interest in electrification. Why would they? Petroleum power had made them kings. Even when they had no choice but to address the environmental consequences of their products, they put far more effort into blocking the Clean Air Act than they did developing the technology to comply with it. It's not surprising that an industry that went to the mat to avoid putting catalytic converters on its cars was unserious about building electric cars.

Nonetheless, beginning in 1966, Detroit did produce a few electric prototypes. General Motors produced the Chevrolet Electrovair, a Corvair converted to run on expensive and impractical silver-zinc batteries. GM followed the Electrovair with a line of hilariously podlike experimental city cars, one of them a gas-electric hybrid and one a pure EV powered by twelve lead-acid batteries. That one, the XP512E, looked like a clown-car version of an AMC Pacer. In 1967, Ford of Britain's research staff revealed its tiny electric city car, the eighty-inch-long Comuta, which never made it into production. American Motors put more effort than its competitors into sexing up its prototype. The AMC Amitron looks like the swinging London vision of the car of the future. Interestingly, the Amitron was designed to run in part on an early lithium battery chemistry, lithium nickel fluoride.

After the clean-transportation push of the 1960s and 1970s ceased, the next opportunity for an electric-car revival came in September 1990, when the California Air Resources Board (or CARB, the agency created in 1967 as a result of the air-quality wars) mandated that automakers make 2 percent of the cars they sold in the state emission-free by 1998. That number would rise to 5 percent in 2001 and rocket to 10 percent in 2003.

The regulations were a direct response to General Motors CEO Roger Smith's announcement on Earth Day of that year that the company was putting its Impact electric-car concept—as the EV1 was then known—into production. The Impact concept car had grown out of the Sunraycer solar-powered car, the product of a collaboration of Aero-Vironment (the Los Angeles company behind the Gossamer Condor, the first human-powered aircraft), GM, and Hughes Aircraft. Sunraycer won the 1987 World Solar Challenge, a highly publicized race across Australia. It was a public-relations coup for GM, and on that success the three companies built the unfortunately named Impact, which Smith first unveiled at the 1990 Los Angeles Auto Show.

The Mandate—it's become a proper noun—spawned the doomed electric-car boomlet of the 1990s. GM and its competitors produced small numbers of electric cars in order to continue doing business in California, the largest market for automobiles in the United States. But although its own CEO inspired the Mandate, GM complied only grudgingly. The company clearly hated doing a car program on someone else's terms. GM repeatedly said that there would be no market for electric cars—that nobody wanted them—and they continued saying this even after events proved them wrong. When GM announced a small test-fleet program called PrEView, in which fifty Impacts were loaned to volunteer drivers for two weeks in order to collect data on the car's performance, they expected a weak response. Instead, they had to close the phone lines. Some twenty-four thousand people called asking to participate in the program from Los Angeles and New York alone.

It's not exactly trafficking in conspiracy theory to say that GM undermined its electric car from the start. This oft-quoted *New York Times* article says quite a bit:

> General Motors is preparing to put its electric vehicle act on the road, and planning for a flop. With pride and pessimism, the company, the furthest along of the Big Three in designing a mass-market electric car, says . . . it has done its best but that the vehicle has come up short . . . Now it hopes that lawmakers and regulators will agree with it and postpone or scrap the deadline.

Regulators eventually postponed the deadline, but they didn't scrap it. Still, it's a miracle that the EV1 was built. As Michael Shnayerson tells the story in his book *The Car That Could*, the EV1 project nearly had its legs cut off numerous times. The small team of engineers responsible for the EV1 were as passionate about the car as GM executives were ambivalent, however, and they scrounged for resources and struggled until they made the car a reality.

When the EV1 was launched in 1996, it was available only for lease, and only in Los Angeles, Phoenix, and Tucson, though eventually San Francisco and Sacramento would get EV1 programs too. Lease rates were generally $499 a month at the car's launch, but that dropped to $349 the following spring. To deal with the quirks of leasing an electric car—making sure customers understood range limitations, holding their hands through the process of installing charging stations in their garages—GM hired a team of specialized salespeople called EV specialists and stationed them at Saturn dealerships in the cities where the EV1 was available. That first generation ran on more than a thousand pounds of lead-acid batteries built by Delphi and got 70–140 miles to a charge, mostly because of its incredible aerodynamic efficiency. GM claimed that the car's drag coefficient of 0.19 made it as aerodynamic as an F-16 fighter jet.

EV1 drivers adored the car. It was fast, smooth, stylish, and silent.

Brand manager Ken Stewart said that EV1 customers were demonstrating a "wonderfully maniacal loyalty." Francis Ford Coppola, Mel Gibson, and other celebrities leased the car. But while GM was rolling in positive press as a result of the EV1, it was simultaneously lobbying the state of California to dismantle the CARB regulations. The company argued that this legislation was forcing a private enterprise to produce, at enormous expense, a product that nobody wanted. It declared that the best alternative to petroleum was not electricity but hydrogen, and if CARB would just give the automakers time, they would usher in a hydrogen-powered future. To prove that no one wanted electric cars, automakers relied on studies like one conducted by the Berkeley professor Kenneth Train. During CARB hearings in 2000, Train said that his research showed that the only way Toyota could get people to "buy" the electric version of its RAV4 would be to "give the average consumer a free RAV4-EV plus a check for approximately $7,000." In reply, the California Electric Transportation Coalition commissioned a study that found that the potential electric-car market in California was actually something like 12 to 18 percent of new light vehicles, or as many as 226,800 electric cars a year. Starting in 2002, Toyota put electric RAV4s on the retail market with a manufacturer's suggested price of $42,510, which came down to $29,510 after rebates. Toyota received more orders than it had cars and ended up building additional vehicles to fill all 328 of them. Then it quietly scrapped the program.

General Motors and DaimlerChrysler eventually succeeded in beating CARB down. GM built the last EV1s in 1999; that generation used nickel-metal-hydride batteries and could run up to 150 miles on a charge. In all, GM leased some eight hundred EV1s between 1996 and 1999. The program was discontinued in 2002, and a year later Rick Wagoner officially canceled the program, saying it was a money loser.

Then GM made one of the biggest public-relations mistakes in its history: it recalled the cars. GM couldn't be responsible for maintaining these discontinued cars for the remainder of their fifteen-year warranties, the argument went. You can't expect us to keep the parts around. A number of EV1 drivers pleaded with the company to let them buy their cars in exchange for freeing GM from warranty obligations, but it didn't work. GM rounded up the EV1s, hauled them out into the desert, and crushed them. Unfortunately for GM, much of the recall process was caught on tape, and that tape was edited into *Who Killed the Electric Car?* In 2006, the year the film came out, Rick Wagoner admitted to a reporter that canceling the EV1 had been a serious blunder.

After the Volt announcement, when the electric car once again appeared to have a chance, those who were there for the EV1 saga insisted that this time, things were different. "In the 1990s we had $20-a-barrel oil," said Ed Kjaer, director of Southern California Edison's Electric Transportation Department. "Fuel economy was thirty-seventh on the hierarchy of importance for purchase decision. We didn't have 9/11, we didn't have global terrorism, and we didn't have—and this is the eight-hundred-pounder—China and India. And we didn't have global warming, in terms of increasing regulatory pressure." What they did have, and what generated the EV1 and its cousins, was air-quality regulation. Once the automakers succeeded in softening the regulation, what motivation did they have to keep making electric cars?

Engineers who worked on both the EV1 and the Volt argued that the EV1's fate portended nothing for the Volt. "It was a different proposition for the EV1," said Jon Bereisa, chief of engineering on the EV1 and an early member of the Volt team. "The battery technology was not there and we knew it, but we believed that we could make up for it by designing a highly efficient vehicle. So what we did is we set about to make the world's most efficient vehicle." The second-generation EV1 extended the car's range to as much as 160 miles, but, according to Bereisa, the large nickel-metal-hydride batteries that made the additional range possible contained costly materials such as cobalt and vanadium and were, as a result, outrageously expensive—as much as $40,000 or $50,000 a battery. Bereisa estimates that GM lost $1 billion on the EV1 project. "We established technical feasibility," he says. "You could say we nailed it to the wall. But we really did not achieve commercial viability."

The first step toward making the Chevrolet Volt commercially viable was filling the giant blank spot at the heart of the car—developing the battery.

Before the concept was even announced, GM had approached General Electric about supplying the batteries for the Volt. The idea was two American icons coming together to build the car of the future. But it wasn't to be. "We were hoping that we could in fact announce some sort of partnership with GE when we showed the Volt," Bob Lutz said, "and as it turned out we couldn't. They were not ready to commit to lithium-ion production."

Instead, GM began searching the globe for potential partners. "We went into a very profound analysis of the top twelve or fourteen lithium-ion battery producers of the world and we assessed them on such things as size, technological capability, reliability in the field, level of automation in their plants, suitability of the chemistry," Lutz said.

The month after the reveal of the concept car, delegations from eight battery manufacturers began filing one after another, props and proposals in hand, into the massive glass-walled Vehicle Engineering Center on GM's sprawling, Eero Saarinen–designed Tech Center campus in Warren, Michigan. Start-ups and multinational giants alike, these companies had survived the initial cut in the Chevy Volt battery-supplier derby, in which some twenty employees from various GM divisions spent two months scrutinizing proposals. They graded each company's batteries on energy and power density, temperature performance, safety, life span, and cost. They weighted each metric by importance and factored in what Tony Posawatz diplomatically called "qualitative factors," such as, Are we going to hate working with these guys?

Each supplier had to prove that its product could store 16 kilowatt-hours of energy, drive the Volt forty miles on electricity alone, launch the car from zero to sixty in eight seconds, run for at least ten years, withstand five thousand full discharges, lose no more than 10 percent of its charge capacity along the way, fit into the tunnel that houses a conventional car's driveshaft, weigh no more than four hundred pounds, and cost as little as possible. And never, ever explode.

Any manufacturer that relied on lithium-cobalt-oxide batteries, which by then were being used in billions of laptops and cell phones, would have to stage a particularly persuasive show. Yes, Tesla was using them, but of the several varieties of lithium-ion batteries on the market or in development, it had recently become clear that lithium-cobalt-oxide batteries were the most prone to the chemical reactions that cause what engineers euphemistically call "thermal runaway." And in early 2007, just months after Sony lithium-ion batteries around the world started going up in flames, this fact was particularly raw.

The crisis began in December 2005, when laptops powered by Sony lithium-ion batteries began catching fire. Videotaped and YouTubed laptop fires began to make the news. In June 2006, images of a laptop bursting into spectacular flames at an Osaka, Japan, business conference circulated on the Internet. The following month a UPS cargo plane was engulfed by fire at the Philadelphia International Airport, and lithium-ion batteries were immediately suspected. That same month, a Sony battery inside a Dell laptop caught fire in a Nevada man's truck, triggering ammunition he had stored in the glove box, igniting the gas tank, and blowing the truck to pieces. On August 14, Dell recalled 4.1 million laptops in what *The New York Times*, citing the Consumer Products Safety Commission, called "the largest safety recall in the history of the consumer electronics industry." The battery disaster made its way into the casual tech argot of the day, as in this blurb for a new flame-retardant laptop cover published on the blog Engadget that September: "We're not sure if these new fire-retardant covers are meant to protect nearby objects in the event of battery explosion, or if they're meant to protect the MacBook from thermal disaster in its surrounding environment—but either way, they're a pretty stylish new necessity." By October, Sony had recalled nearly ten million batteries worldwide. The explanation appeared to be that in a certain batch of batteries manufactured at Sony's Fukushima plant, metal fragments made it into the electrolyte during the process of crimping shut the battery's metal shell. Eventually, in some cases, those metal fragments caused a short circuit; in the worst cases, that led to a fireworks display and a viral media phenomenon. The episode made clear the risks of packing so much energy into such a small container. If something goes wrong, there's a lot of energy to be released.

So the Volt-battery-supplier hunt took place in a sensitive context. Meanwhile, Bob Boniface's team was busy with the production design for the Volt—a project he hadn't necessarily expected to still be working on. "I didn't think it was gonna go over that well," he said. "There was an outpouring of goodwill. I miscalculated."

Soon there were signs that GM felt cornered by the Volt—that it knew it might have on its hands, but that it was also afraid the car had the potential to backfire in a devastating way. On March 23, 2007, an article appeared in *The Detroit News* headlined "GM Tries to Unplug Volt Hype." According to the piece, the fact that the Volt was based on batteries that the company didn't yet have in hand "led to intense debate within GM over whether it was wise to show the Volt in Detroit. And now that the world's waiting for GM to deliver what could be the biggest environmental breakthrough so far this century, company officials are actively trying to temper expectations." The article tells of a GM-initiated background session in which journalists were reminded of the many technological issues that, taken together, meant that the car might never actually reach the road. "The pressure is intense," said Nick Zielenski, chief engineer on the Volt. "We came out with this idea and now people are saying, 'Okay, where is this car? We want it now.'"

While the expectations management happened on one front, the scramble to build the Volt continued on many others, and in the battery-supplier competition, the finalists soon became clear. As Lance Turner, lead engineer for battery development, put it, "It's very easy to deal with someone close by"—that would be Compact Power, Inc., whose Troy, Michigan, offices are a fifteen-minute drive from GM's Warren battery lab—"and it's very easy to deal with someone we've dealt with before." That would be Continental, the German auto parts manufacturer responsible for bundling battery cells from a young Boston-area start-up called A123 Systems into a functional battery pack.

In addition to pitting company against company, the Volt battery competition would be a contest between two competing strains of lithium-ion battery chemistry: lithium manganese oxide, which Compact Power used, and lithium iron phosphate, which A123 built its company on. Those two chemistries differed greatly from one another and from the lithium-cobalt-oxide batteries used in consumer electronics.

Because both of them did away with expensive, toxic cobalt, both were potentially cheaper and better for the environment than lithium cobalt oxide. They were also safer, and that was A123's strongest selling point: the double covalent bonds that held the phosphate group together were the strongest bonds in nature, and that made it extremely difficult for that chemistry to react inappropriately. The trade-off for greater inherent stability, however, was lower energy density than its competitor from Compact Power.

By the company's annual shareholder meeting in June, Bob Lutz was ready to announce the finalists. There was a certain David and Goliath dynamic in the matchup. Compact Power (often called CPI) is the frontier settlement of one of the largest consumer-electronics manufacturers on the planet—the Korean company LG Chem. Behind CPI's modest Troy headquarters lies the unseen power and weight of a company that builds tens of millions of lithium-ion batteries each year. Behind A123 was buzz. By 2007, the press had become infatuated with A123, the company that MIT professor Yet-Ming Chiang and three colleagues had founded six years earlier. They were made for the media, a clean-energy throwback to the hip start-ups of the dot-com era, and flattering profile after profile portrayed the young company as the very picture of American high-tech ingenuity.

CPI had been working for five years on the lithium-manganese-oxide chemistry that John Goodenough and Michael Thackeray first developed at Oxford in the 1980s. It was attractive for electric cars because of the low cost of manganese and because of its inherent power, which is best understood through an analogy: Energy is how much water fits in a bottle; power is how quickly you can pour it out. In a car, power equals acceleration.

Energy capacity was another major concern. "We had to have a cell that effectively doubled the energy capacity of a typical hybrid cell," said Prabakhar Patil, the CEO of Compact Power. CPI's seventy staffers worked nights and weekends for four months after the shareholder meeting. They then surprised the engineers in GM's battery lab by delivering their first finished battery pack right on time, on Halloween day.

At the same time that GM technicians worked eagerly to hook the CPI battery to the machines that would test its worth, and at the same time the local papers were gleefully reporting on the first battery's arrival in Warren, A123's first pack was stuck in Customs. The U.S. Department of Transportation considers lithium-ion batteries dangerous material, which made it difficult to get the pack delivered from the packaging facilities in Germany. It probably didn't help that the stainless-steel casing wrapped around A123's cells made the battery look like a nuclear weapon from a Jerry Bruckheimer movie.

After more than two months of ulcerous delay, in January 2008, Customs released the batteries. Jon Lauckner was in Washington, D.C., sitting on a panel on plug-in hybrids at the Center for American Progress, and he insisted he be told the minute the battery reached the lab. Lauckner took the stage and began to field questions from the audience. "Where are you with the second battery?" someone asked. Lauckner looked down at his BlackBerry and replied, "It arrived in our lab five minutes ago."

The batteries that both companies submitted weighed approximately four hundred pounds and, stood on end, reached a height of six feet. Each $10,000-plus, T-shaped monolith contained more than two hundred individual 3.6-volt lithium-ion cells, bundled together in groups of three, then wired in series and kept from overheating by an elaborate liquid cooling mechanism. A computerized monitoring system inside each battery pack conducted this little orchestra, coordinating the actions of the individual cells, balancing voltage, and watching, above all, for any indication that a cell might be failing, shorting out, or otherwise threatening the stability of the system. The batteries were engineered to propel the 3,520-pound Volt forty miles. (To make sure the battery lasts the warranty-required ten years and 150,000 miles, the Volt team initially decided to use only half of the 16 kilowatt-hours, never charging it above 80 percent of capacity and never depleting it below 30 percent, thereby reducing the chemical strain on the battery's cells.)

By the summer of 2008, those batteries were powering the earliest Volt prototypes around GM's proving grounds in Milford, Michigan. These were the Mali-Volts—Chevy Malibus gutted and fitted with the

Volt power train. In early June, Lutz described driving the Volt as both thrilling and eerie. "It's like being in a conventional car at seventy miles an hour and coasting with no engine," he told a reporter for Greenfuelsforecast.com. Back in the lab, he reported, the batteries were performing well. Some of the welds that tie the individual battery cells together had failed, but that was expected and not fundamentally a big deal; the team was increasingly confident. "The guys are now convinced that unless we have some sudden whoops! that we don't see, we're good for November 2010," he said.

Such a showstopper, if it did appear, would most likely show in longevity tests at GM's Warren battery lab. There, those celebrated first two battery packs had, since their arrival, been constantly subjected to the abuses of the pack cycler, a refrigerator-size device that tests cycle life—how many times the battery can be discharged and charged again without deterioration—and another apparatus called the thermal chamber. In two years on the pack cycler, engineers can put a battery through the equivalent of 150,000 real-world miles. The only way to see how a battery ages over ten years, though, is to make the battery, use it for ten years, and see what happens—unless you have only two years, as GM did. Then you artificially accelerate the aging process by heating the batteries in a giant metal sauna for months on end. The first Volt batteries were scheduled to hit the crucial ten-year mark in April 2010—a scant seven months before the Volt was set to go into production.

The general vibe at an industry convention may or may not be a valid metric for measuring the momentum behind a new technology. Nonetheless, at the Plug-In conference in July 2008, it became clear that even the most gun-shy members of the electric-car activist scene, people for whom the failure of the EV1 remained an exposed nerve ending, were allowing themselves to feel optimistic. This time, so many factors were aligned—$4-a-gallon gas, growing awareness of global warming, the desperation of the Big Three automakers, the major advance of the lithium-ion battery. "I've worked in the battery business for twenty-nine years," Michael Andrew, a project manager at Johnson Controls–

Saft, a joint venture between the Milwaukee-based auto-parts supplier Johnson Controls and the French battery company Saft, told the audience during a preconference workshop on lithium-ion batteries. "What's changed to cause so much optimism? This is it."

That didn't mean everything was going smoothly. Tesla had had a rough couple of years since unveiling the Roadster. Depending on whom you ask, the problem was either Martin Eberhard, Elon Musk, or both. According to Eberhard, Musk was a disruptive force, intruding on the design process, creating delays, and driving up costs by changing the headlights at a cost of $500,000, redesigning the chassis to lower the door sill by two inches (his wife had trouble getting out of the car; cost: $2 million), ordering custom seats ($1 million) and insisting on a new carbon-fiber body rather than the fiberglass panels used originally. According to Musk, Eberhard had proved a disastrous CEO, which is why in 2007 he was demoted and then pushed out of the company. When Musk told his side of the story to *Fortune* that July, he said the only reason he had kept quiet so far about the conflict was that he "was too busy trying to fix the fucking mess [Eberhard] left. I will say, I have never met someone who is as capable of creating such a disinformation campaign as Martin Eberhard."

True, Eberhard had been fairly belligerent in telling his side of the story. His rants on his Tesla Founder's blog became famous for their bile. By the time this book was written, the blog, teslafounders.com, had been deleted, but in July 2008 *Fortune* quoted a "typical" Eberhard post: "The company has changed so tremendously since I started. It's very secretive and cold now. It's like they're trying to root out and destroy any of its heart that might still be beating."

In any case, Tesla had blown through its launch date of August 2007. By the time the Plug-In conference began in San Jose, a quick drive down the freeway from Tesla headquarters in San Carlos, Tesla had delivered only seven "Founder's Series" cars, to company intimates and big-time investors like Larry Page and Sergey Brin, along with Musk and Eberhard. Tesla's strife drew its own spectators, and blogs like the Silicon Valley–oriented gossip site Valleywag.com began obsessively covering the company's every misstep.

Meanwhile, as GM's fortunes declined, the Volt's critics were relentless. A month after the Plug-In conference, Bill Reinert, manager of advanced technology at Toyota, told *EV World* that Toyota employees had a "death watch" going for the Tesla Model S, Fisker Karma, and Chevy Volt. The reason: the outrageous cost of the batteries. A public-relations rep attempted to walk back Reinert's comments—"As a company, we do not have an official death watch anywhere," she told the website Greentechmedia.com—but the sentiment was consistent with Toyota's stance on the Volt since the car was revealed.

At Plug-In, GM had dispatched Jon Lauckner to preach to the converted and fight back the hordes who believed the Volt was nothing more than a wall of smoke and mirrors designed to distract the world from GM's financial catastrophe. Over breakfast the first day of the conference, he smiled mischievously when I asked him about the project's many doubters. "The window's closing on the skeptics," he said. "And the only thing that's going to be left at the end of the day is: Are we on time?"

The next day Lauckner defended the Volt in a conference room the size of a football field, before a crowd of nearly seven hundred. He said he had planned on giving a general update on the Volt project and announcing a partnership GM was starting with a few dozen North American electrical utilities, but that changed, because there was a fresh piece of scathing criticism to respond to. Three weeks earlier, an op-ed in *The Wall Street Journal* had accused the Volt of being nothing more than a ploy to make the federal government feel a little fuzzier about bailing out GM. Headline: "What Is GM Thinking?" Lauckner had read the column while in China on business, and he was livid. He quickly e-mailed an angry rebuttal to several top executives and PR people marked "high" importance. "I won't deny this raised my blood pressure," he told the audience. "It's absolute nonsense. It called into question whether the search for automotive technology that doesn't involve petroleum is worthwhile." The first PowerPoint slide to appear on the projection screen behind Lauckner showed the headline from *The Wall Street Journal* with a new byline: Jon J. Lauckner. The gag drew tepid laughter.

"Pick your issue, and the common denominator is oil," he said. "One fact stands out above all others: Going forward, we can no longer rely solely on oil to supply auto energy requirements." Which alternative energy source is the answer, well, that's an open question. "However, we increasingly believe the solution involves the electrification of the automobile as soon as possible. There has been a shift in the debate from 'if' to 'when.'"

Then, for some reason, he decided to probe at the old EV1 wound—old to him, perhaps, but still raw in this audience's mind. "Some folks have recently suggested that we just dust off the tooling—if we can still find it—and crank up production of the EV1," he said. "And look: the technology of the EV1 was state of the art ten years ago. But GM has chosen to put our efforts behind newer and better technology that will have greater functionality and therefore a much greater chance of high-volume marketplace acceptance."

That technology, obviously, was lithium ion, and when the Volt was unveiled a year and a half earlier, "critics voiced doubts," Lauckner acknowledged. "Lithium ion was a dream. Even if it was achievable, we couldn't bring it to market by 2010. Today I can tell you with a lot more development under our belts that we're confident."

6

THE LITHIUM WARS

On a damp morning in November 2009, Yet-Ming Chiang was in prime form. In an expansive conference room in Boston's Back Bay neighborhood, he was addressing the hive mind of the advanced battery research community—an audience of more than a hundred academic and industry researchers gathered for the fall meeting of the Materials Research Society. His half-hour talk was a tour de force of scientific pitchmanship. The fifty-one-year-old scientist was boyishly energetic, with the poise, confidence, and polish of a man absolutely sure of his work, a man who might be on the verge of making a seriously large amount of money.

He sprinted through a series of slides that demonstrated how the exotic electrode powders that he and his underlings spent their time synthesizing in a lab across the Charles River scaled into the technology that could power a new generation of hybrid and electric vehicles. He explained how earlier that year, his company, A123 Systems, had installed the "highest powered lithium-ion battery available today" in a McLaren Mercedes Formula One "Kinetic Energy Recovery System"– class race car, and how in a subsequent race the extraordinary bursts of power that the battery delivered on demand helped the driver move from eighteenth out of twenty to finish in fourth place. He explained that today's lithium-ion batteries are "highly mass and volume inefficient," and that by mass more than half of the stuff in a battery was inactive, auxiliary material, electrochemical packing peanuts that Chiang would very much like to replace with more active electrode material. We can improve lithium-ion batteries by a factor of two just by using the space inside them more efficiently, he said. Thicker, denser electrodes, like the ones being developed in his lab, could help accomplish just that.

He didn't mention that A123 Systems is built on a disputed technology, the subject of interminable legal wrangling and scientific controversy. He didn't mention the Canadian company that claims to have the exclusive right to manufacture and sell the kind of batteries that had made A123 one of the most famous among the new crop of American clean-tech start-ups. He didn't have to, of course. Everyone in the room knew the story, or at least one side of it, quite well.

Chiang also chaired the final session of the day, and afterward I hung around until the well-wishers and graduate students eager to introduce themselves to him had dispersed. We had been in touch by e-mail, and I walked to the front of the room to introduce myself. He asked me to remind him what, exactly, I was researching, and as soon as I got three words out he cut me off: "The lithium wars of the early twenty-first century!" he said with a grin. "I tell my students they'll look back and be able to say, 'I was there.'"

A central battle in the lithium wars began in the early 1990s, a few years after John Goodenough left Oxford to take a post at the University of Texas at Austin. Sony's commercialization of the lithium-ion battery had cemented John Goodenough's reputation as the leader of his field. He may not have seen a penny in royalties, but in solid-state chemistry circles he had become well-known as the man whose compound had transformed portable electronics. In 1986 he had brought a postdoc over from Oxford named Arumugam Manthiram, and together they had established a solid-state chemistry lab in UT's sprawling engineering school. In the late 1980s, during the high-temperature superconductor craze, their battery research faded in priority, but the Sony announcement retrained the scientific community's attention on the subject. Soon scientists were searching for a cheaper and safer successor to lithium cobalt oxide, one that replaced an expensive, toxic material (cobalt) with a cheap, abundant, benign element like iron.

In 1993, a visiting scientist named Shigeto Okada arrived in Goodenough's lab. He was a researcher at Nippon Telegraph and Telephone, otherwise known as NTT. Goodenough put him to work with two other scientists, a graduate student named Akshaya Padhi and a postdoc named Kirakodu Nanjundaswamy, studying a few variations on an iron-based compound Goodenough had worked on previously. "It was supposed to be a perfectly simple scientific study, not necessarily aimed at doing a battery at that point," Goodenough said.

At the end of the spring semester, Goodenough and his wife left for a sojourn at their house in New Hampshire, and while they were away Padhi spent his time studying a synthetic version of the mineral triphylite, or lithium iron phosphate. Surprisingly, it showed promise for use as a battery electrode. He had some luck putting it through "reversible" intercalation reactions—the atomic-scale burrowing that happens when lithium ions swim over to the electrode, dig inside, and dock there until the battery is recharged—and this meant that it was stable. The new compound also had the benefit of being made of nothing but cheap, almost free elements, which was exactly the kind of thing they needed to succeed the lithium cobalt oxide that Sony was now selling to the world. There was plenty of work left to do, however, and in the end, the results of their study were anemic. The new compound had a low capacity and was terrible at conducting electrons. Ions, no problem. Ions flew through this compound. Electrons were a different story, however, and a battery terminal in which electrons get bogged down as if in quicksand is useless. Nonetheless, the results were interesting enough that in 1996, Goodenough and Padhi decided to present their results at a meeting of the Electrochemical Society in Los Angeles.

Michel Armand wasn't planning to attend the conference, but when he saw the abstract for Goodenough's paper, he knew he had to go to Austin. By then Armand was a visiting professor at the University of Montreal. He was also consulting for the energy company Hydro-Québec, which since 1978 had been doing R & D on a novel battery Armand invented in the early 1970s—a cell that used metallic lithium for the anode and a solid polymer to act as both the separator and the electro-lyte. After leaving Bob Huggins's lab at Stanford, he had gone back to France, and while there he had attempted to make iron phosphate into a lithium battery material, suspecting that it would make a good positive electrode for his polymer batteries. For technical reasons, he was never even able to synthesize the material, which is why when Good-enough announced that he and a student had made a batch and managed to get some interesting results, the fact that it appeared to be a relatively limp battery material didn't matter to Armand. It existed, and worked at least a little bit. Armand was obsessed with the possibility of making battery electrodes based on cheap, almost infinitely available iron. "I immediately took the first flight out," he said.

Armand arrived in Austin with a small entourage to secure rights to lithium iron phosphate for Hydro-Québec. "He was very reluctant, because he didn't believe in his compound," Armand says of Goodenough. The trip was a success, however, and soon Hydro-Québec had an exclusive license on the technology, which meant that only Hydro-Québec—or a company that Hydro-Québec licensed the rights to—could legally manufacture and sell lithium iron phosphate electrode powder in North America.

Within six months, Armand thought he had learned how to make the compound work. He believed that making particles of lithium iron phosphate that were each about the size of a particle of soot could solve the problem of low electronic conductivity. When individual particles "go nano," or get down to the unfathomably tiny scale of less than a hundred nanometers wide, the particles are almost all surface area, and more surface area allows electrons to roam more freely. In the process of making those small particles, however, Armand's group happened upon the second key to making lithium iron phosphate work. They started with a precursor material made of iron, phosphorus, and oxygen. Then they added a lithium compound and fired it. The burning of the lithium-containing compound ended up coating the tiny particles with carbon, and the conductivity shot up. "It solved everything," Armand said. "The phosphate was perfect."

As it turned out, Goodenough, his student Padhi, and then Armand had developed something significant, a substance that would go on to be called one of the greatest materials-science advances of the decade. "But," Armand says, "that was also the beginning of what would be—will remain—the biggest scandal in lithium batteries."

A few years after Michel Armand's sprint to Austin, Yet-Ming Chiang's group began working on "self-assembling" batteries, a far-horizon concept that means exactly what it sounds like. "We were trying to design into [different materials] the necessary attractive and repulsive forces to have a system in which cathode and anode particles assembled themselves," Chiang said. To do so they were studying olivines, the class of compounds that includes lithium iron phosphate, which had properties that made them good candidates for that experiment. Chiang and his team were "doping" various olivines in an attempt to make them better conductors of electrons. Doping—adding tiny, targeted dashes of impurities to a material in order to tweak its electronic structure and therefore change its behavior—is one way materials scientists bend nature to their will. It's how scientists can engineer the interior organization and behavior of electrons and atoms until, for example, what was a wafer of plain silicon becomes the basis for a microchip. When Chiang's student Sung-Yoon Chung applied this technique to lithium iron phosphate, embedding niobium or zirconium atoms in just the right spots in the crystalline lattice, it seemed to cause an astonishing increase in the ability of the material to conduct electricity. It was like turning salt into metal. These were "very surprising results," Chiang said.

In October 2002, Chiang's group published a paper that presented doped lithium iron phosphate as the next great hope for hybrid and electric vehicles. The paper argued that they had improved it on the atomic scale in such a way that it could make a battery cathode that could be completely discharged in three minutes, which is the kind of raw power that an electric-car battery needs. It was a breakthrough, Chiang's paper argued, that "may allow development of lithium batteries with the highest power density yet."

Goodenough's old collaborator Michael Thackeray, who was by then working at Argonne National Laboratory, wrote an accompanying editorial that emphasized the potential significance of Chiang's experiment.

This had "exciting implications" for "a new generation of lithium-ion batteries." Thackeray acknowledged "one slightly controversial aspect" of the research: "that the olivine powders were synthesized from carbon containing precursors . . . Carbon can, of course, contribute significantly to electronic conductivity. Nevertheless, Chiang and colleagues carefully addressed this possibility and ruled it out." His conclusion: "These results will spark much interest in the lithium battery community, who will undoubtedly want to repeat the experiments quickly to verify these very significant increases in electronic conductivity."

That was a bit of an understatement. The idea that adding a small number of metal atoms to lithium iron phosphate could transform it into a good electronic conductor generated considerable skepticism in the lithium-ion research community. To many, it just didn't seem possible to transform this material so greatly with such a small tweak to its chemical composition.

Michel Armand was incensed when he saw the paper. He believed there was no way Chiang's method could have worked. To Armand, it was clear that Chiang had done essentially the same thing he had some years earlier—that in the process of preparing the material, he had unwittingly coated the particles with carbon. Instead of accepting that a carbon coat had made the material usable, however—which Armand had already demonstrated and presented publicly—Chiang simply clung to the more remarkable and useful explanation. Chiang's paper was "false science," Armand told me, something that should have never gotten past the independent scientists who judge studies before they can be published in a peer-reviewed journal like *Nature Materials*.

"When the paper came out, I wrote to the editor and said, 'I'm very sorry for you, because you've got a big scientific goof,'" Armand said. Armand began reproducing the experiments in Chiang's paper, and soon he had formulated a rebuttal. Authored by Armand and two colleagues, the response was published as a letter to the editor in the 2003 issue of *Nature Materials*. The retort was delivered in the understated smack-talk of a scientific journal. "We suggest that the effects seen by Chung *et al.*"—Chiang's student Sung-Yoon Chung was the first author on the paper—"are due to carbon for low-temperature samples, and to low-valency iron derivatives . . . It is beyond, not the scope, but the length of this letter to discuss the juggling of point defect chemistry equations to justify the results. . . . Unambiguously, it is the polyolefin worn from jars, subsequently charred into carbon, which is responsible for the good use of the LiFePO$_4$ electrode."

To translate, Armand was accusing Chiang of either misunderstanding or misrepresenting his research. Chiang hadn't doped anything, Armand argued. Instead, some of the lining from jars used in the experiment had been charred into carbon in the process of synthesizing the material, and that carbon had then coated the particles. There was a second fluke at work too, Armand argued. A metallic compound of iron and phosphorus (Fe$_2$P) had also coated the particles, making it easier still for electrons to move around in the material. Together, these two lucky accidents made LiFePO$_4$ into a fierce conductor of electrons.

Immediately following the Armand group's response came a rebuttal from Chiang's team. "The 'reproduced' experiments by Ravet et al."—the first name on the Armand group's letter was Nathalie Ravet, a chemist at the University of Montreal—"have, on closer reading, clear differences in procedure from ours." Chiang's team argued that every objection Armand raised had been addressed in the original paper. They had accounted for and isolated the effect of extraneous carbon. Chiang's team also argued that Armand's response was fundamentally flawed. Years later, Chiang told me, "Today there's absolutely no doubt that [lithium iron phosphate] can be doped, and it's not only our work but other published work that shows that." On the matter of the extraneous carbon, he said, "In that first paper we measured and calculated the amount of carbon and showed that results were not correlated with carbon."

In February 2004, an independent party entered the fray. Linda Nazar, a professor of chemistry at the University of Waterloo, had been studying lithium iron phosphate since 2000. After reproducing Chiang's experiments, she submitted her paper to *Nature Materials*. As she put it, "The description of a highly electronically conductive phosphate challenges conventional wisdom."

There were stakes beyond scientific esteem. By the time his 2002 paper was published, Yet-Ming Chiang and Bart Riley, Ric Fulop, and David Vieau were already in the process of spinning out Chiang's research on doped lithium iron phosphate into A123 Systems, named after the force constant used in the study of nanoparticles. They started with private capital and a $100,000 small-business grant from the Department of Energy. Then in late 2003, the company got its first big break—a contract to supply power-tool batteries to Black & Decker.

Nazar's paper was published in 2004. She concluded that it wasn't the carbon alone in Chiang's material that provided the extraordinary increase in electronic conductivity, but it also wasn't Chiang's dopant. Instead it was a coating of a highly conductive iron phosphide, an unexpected contaminant that Nazar and her collaborators observed directly using a transmission electron microscope. According to Armand, Nazar's paper showed that "there was nothing new in [Chiang's] phosphate."

Chiang had a response for this too: "We also published a paper around that same time in *Electrochemical and Solid-State Letters* that showed that the materials we were measuring had discrete particles of metal-rich phosphides, not the continuous metal-rich phosphides necessary to form a conductive path." Chiang argued that continuous streams appeared in Nazar's experiment because she used different gases in her experiment, which created an environment that allowed them to form.

Under normal circumstances, a dispute this arcane would have never spilled out of the scientific journals. But the commercial promise of lithium iron phosphate was clear to everyone involved. Chiang had a company on the line. Soon, a parallel fight began in court.

In November 2005, A123 had its coming-out party. The company had been operating in stealth mode for four years, quietly raising money, looking for customers. In that time, A123 had raised more than $30 million in venture capital from pedigreed investors such as Qualcomm, Motorola, and Sequoia Capital, so when A123 started talking about its plans for the future, the press was eager to listen. *The Wall Street Journal*, the magazine *Red Herring*, and a regional tech-industry publication all reported on the company that fall. Chiang, who had previous experience with a high-tech start-up, was a natural salesman. "We think this is equivalent to the impact lithium-ion batteries originally had on the electronics industry in the mid-1990s, where it unseated nickel-metal-hydride batteries as the standard," he told a reporter. A123's 36-volt "nano phosphate" battery packs were scheduled to start selling the following summer, in Black & Decker's DeWALT line of high-end power tools. Due to their ability to dump electricity rapidly, the batteries would soon be powering a series of saws, a hammer drill, and an impact wrench.

Hydro-Québec sent A123 a warning in late 2005, a letter accusing them of violating Hydro-Québec's exclusive license on U.S. patents 5,910,382 and 6,514,640, which the University of Texas held on Goodenough's lithium iron phosphate technology. The letter put A123 on notice: if they didn't stop building lithium iron phosphate batteries right away, they could expect a lawsuit.

But A123 struck first. On April 7, 2006, the company filed an action seeking declaratory judgment against Hydro-Québec, arguing that "neither the lithium metal phosphate technology nor any other product made, used, or sold by A123 infringes" on either patent, according to a complaint filed with the U.S. District Court of Massachusetts. On September 8, they requested a reexamination of the patents, arguing that they overlapped with several Japanese patents that were filed earlier.

Three days later, the fight came to a very public head. The University of Texas stepped in and, along with Hydro-Québec, sued everyone involved in manufacturing and marketing A123's debut power-tool batteries: A123, Black & Decker, and China BAK Battery. "Nearly a decade ago," read an *Austin American-Statesman* article on the legal battle, "the University of Texas licensed two patents that were supposed to help power the next generation of laptop computers, cell phones and other staples of the tech age. Today, the university says, this longer-lived and more powerful lithium-ion battery is finally hitting the mass market . . . The problem, according to the university, is that Black & Decker is essentially bootlegging its technology."

Litigation, patent disputes, overpromising, and get-rich-quick hype have stained the battery business since its inception. In fact, the earliest attempts to commercialize the rechargeable lead-acid battery were shady enough that in the 1880s the storage battery business was about as reputable as cash-for-gold schemes are today. Back then, the French inventor Camille Faure, who developed a method for the fast manufacture of lead-acid batteries, and a businessman named Gustave Philippart wanted to make the storage battery an indispensable component in the electrical lighting systems that were beginning to emerge. The pitch was based on efficiency and labor costs: With a large array of batteries, employees at a central power station could, during normal working hours, run dynamos to charge the batteries. Then they could go home, and when customers turned on their lights in the evening they would draw electricity from the bank of batteries charged earlier in the day. Almost without exception, however, the battery companies of this era failed, in part because their technology was still relatively crude, but also because they were run by a fractious bunch of hustlers—primarily Philippart, Charles Brush in England, and a character with the perfect evildoer name of E. Volckmar—who drove the industry into a swamp of patent lawsuits, market-cornering schemes, and vicious public arguments. The companies built on Faure's technology "were an unfortunate attempt to make too much money too quickly with too little technology," wrote Richard H. Schallenberg. The public saw men like Philippart as scam artists and put the battery in the same category as snake oil. This is what inspired Thomas Edison to call the battery "a catch-penny, a sensation, a mechanism for swindling the public by stock companies," a corrupting genie's jar that brings out a man's "latent capacity for lying."

So the fight over ownership of lithium iron phosphate had plenty of precedent. Actually, the A123 conflict wasn't even the first legal battle over the compound. That one began in 2001, when the University of Texas and Hydro-Québec filed suit against Nippon Telegraph and Telephone. The coalition had been shopping their technology in Japan when they heard some surprising news. During negotiations with Sony and Matsushita, the North Americans learned that NTT had been granted a Japanese patent on the same substance four years earlier. And they had filed for it in November 1995—thirteen months after Shigeto Okada had returned to Japan from Goodenough's lab in Austin.

Akshaya Padhi had finished his work on lithium iron phosphate in the fall of 1994; Okada returned to Japan on October 9 of that year. Despite Goodenough's warnings, Padhi hadn't exactly been discreet about his findings. "Padhi told [Okada] everything that was going on," Goodenough said. Padhi even continued to e-mail Okada details of his research well after Okada had returned to Japan. In court documents, the University of Texas and Hydro-Québec claim that "upon returning to NTT in Japan, [Okada] disclosed the confidential information, which NTT used to apply for a Japanese patent." Goodenough said that NTT had long wanted to make lithium iron phosphate but had never figured out how to fabricate it properly. That part, Goodenough claims, they learned in Texas, through Okada. And because they had thought of lithium iron phosphate at one point, Goodenough said, they thought they were entitled to a patent.

"As a result," Goodenough said, "we ended up with a lawsuit. And I learned how lawyers work." With that, Goodenough unleashed an epic laugh lasting ten seconds and containing at least thirty separate bellowing, unrestrained chuckles.

In the suit, the university and Hydro-Québec accused NTT of, inter alia, "tortuous interference, unfair competition, misappropriation of trade secrets, conversion, and breach of a confidential relationship." The university's lawyer, a Dallas suit named William Brewer, did some belligerent gloating to the *Houston Chronicle* three years later, in 2004. "I think that when a jury hears the facts, they might just hand me a rope," Brewer puffed.

The most damning piece of evidence was probably a note that Okada had faxed Goodenough from Japan shortly before the plaintiffs filed suit. In it Okada admitted that, upon closer inspection of his notes, maybe things were not quite on the up and up. "It seems to show the themes of Swamy and Padhi as you said," he wrote in imperfect English. "You have taught us them apparently. I have forgotten this discussion in UT up to now. I was unconscious of having made a mistake. I must apologize to you and Padhi again. To take the responsibility of my sin, I send this paper by fax."

After the university filed suit, Okada walked back his alleged confession, and NTT denied any obligation. Of Okada, Brewer told the *Chronicle*, "He is a liar. He is flat out lying and hoping to get away with it."

———

The question at the core of the A123 saga is whether Yet-Ming Chiang transformed the chemical that John Goodenough patented into a new and eminently more useful compound, or whether his compound was essentially the same as what had come before.

A123's position is simple: Their cathode material has a different chemical formula, and therefore is a new invention that is the work of Yet-Ming Chiang and his colleagues.

On the scientific front, Armand's argument didn't depart much over the years from the case he made in the pages of *Nature Materials* in 2003. After that, Linda Nazar became the chief agitator. Nazar and Chiang became well-known debate partners as Nazar continued to challenge Chiang's research on lithium iron phosphate and Chiang continued to respond. In 2006, Nazar published another paper on the subject. Then in 2008, she published a paper that further clarified the role of Chiang's dopant—in a way that was not at all favorable to Chiang. Naturally, Chiang hit back, publishing another paper in 2009. Nazar and her colleagues issued what she said was their final entry in the saga in 2010, and Chiang published a reply in the same journal.

In a phone conversation, Nazar seemed tired of the drama yet unable to let it go. She would say she didn't want to comment on the controversy, and then she would comment anyway. She was adamant that it was a scientific disagreement, not a clash of personalities, and because she has no financial interest in lithium iron phosphate batteries, it's not a business matter. "Scientifically there has been a disagreement on the nature of electronic conductivity enhancement in lithium iron phosphate," she said. "And the science speaks for itself. It shouldn't get down to anything personal between scientists; it's simply based on science. And the scientific community judges for themselves as well." A number of researchers in the field told me they believed the consensus is that Chiang's initial results were not as claimed, but no one was willing to be quoted saying it.

Peter Bruce, who, like Nazar, has served as an expert for Hydro-Québec in their litigation against A123, explained the state of the science, which is still unsettled. "The difficulty in answering your question is that it still today remains a controversial issue. As indeed does the issue of what actually is the limiting factor in lithium iron phosphate. Some people believe it's the electronic, and some people believe it's the ionic conductivity. And if it isn't the electronic conductivity, then it doesn't matter so much if you improve it." In short, "There's still a lot that's not understood about this material."

Nazar is quick to point out that, scientific discussions about *why* Chiang's invention works so well aside, it does work well. "Suffice it to say that he has a successful company," she said. "The materials seem to work electrochemically."

Chiang, for his part, stands by the original results. He also said that in the years since the 2002 paper, he's discovered that the process of doping yields additional benefits, which he has published. "The behavior of these materials turned out to be richer than we had originally thought," he said. "It left a lot of opportunity for further improvements in the material."

While the scientific dispute among Chiang, Nazar, and others played out in the pages of peer-reviewed journals, the patent battles continued. By January 2007, the U.S. Patent and Trademark Office (PTO) had agreed to reexamine both patents, putting the litigation on hold until that process was complete. No injunction had been filed, so A123 was free to keep making batteries. The patent office eventually rejected all the original claims of both patents. In response, the University of Texas then narrowed its claims. Finally, by May 2009, the PTO accepted the amended patents. The lawsuits were then free to move forward.

"It's been all messed up because of these patent disputes, and people are stealing it right and left," Goodenough said. He wasn't referring to A123 as much as the numerous other companies that have jumped into the phosphate game—the Austin-based company Valence, for instance, which supplies lithium-ion batteries to Segway and other customers. Hydro-Québec sued them in 2006; the following year, Valence sued Phostech Lithium. (In February 2011, Valence won the latter suit.) Or BYD, the Chinese battery and electric-car manufacturer that aims to lead the world in electric-car sales by 2015 using its "Fe" battery, "Fe," of course, being the elemental sign for iron. If BYD begins selling those batteries in the United States, it may be inviting a lawsuit. Aside from the Valence-Phostech case, only one other major legal disputes involving lithium iron phosphate had been settled by the time of this book's publication. In October 2008, Nippon Telegraph and Telephone settled its suit out of court. They agreed to pay the plaintiffs $30 million and give them an exclusive license on their patents, while still denying any wrongdoing.

On the case of A123, Goodenough is both harsh and magnanimous—not angry, but disappointed. The money, well, that he'd just like to use to endow a chair for a fellow professor at the University of Texas. On the science, however, he is blunt. "A123 didn't understand chemistry," he said. Nonetheless, Goodenough gives A123 credit for generating a crush of interest in a material that he at first didn't fully believe in. "They got a good material," Goodenough said. "And A123 did a very good marketing job. They are excellent marketers."

Michel Armand is still furious. To Armand, patents "are tearing the community apart." The saga of lithium iron phosphate is a "horror story" of "meanness and greed." "Oh, yeah, of course, he feels very angry," Goodenough said. "Because I must say Michel Armand was the one who recognized that the $LiFePO_4$ was potentially very interesting."

Even before Yet-Ming Chiang published the paper that so profoundly offended him, Armand had in quick succession lost control of the two technologies that he was most passionate about. His saga began when Hydro-Québec transferred to its venture capital arm a subsidiary called Argo-Tech, which had been created to commercialize a form of the lithium-metal-polymer battery that Armand had invented two decades earlier. In 1999 the venture capital arm brought in a new CEO, and Argo-Tech got a new name: Avestor. The company hoped to get lithium-metal-polymer batteries into the automotive and stationary storage markets as soon as possible. After the shakeup, a close colleague of Armand's named Michel Gauthier, who had been program manager for lithium polymer batteries at Argo-Tech, left Hydro-Québec. Then in 2001, Hydro-Québec signed over 50 percent of Avestor to Kerr-McGee Chemical, an Oklahoma-based oil-and-gas company and manufacturer of certain battery compounds. "An oil company!" Armand said. He believes Avestor's fate was sealed as soon as that deal was inked.

That same year, Hydro-Québec transferred its license to manufacture lithium iron phosphate electrode powder to a newly formed Montreal-based company called Phostech Lithium. The president was Michel Gauthier, who between leaving Hydro-Québec in 1999 and the formation of Phostech had spent two years as a visiting researcher in Armand's lab at the University of Montreal.

In September 2002, Avestor began commercial production of 48-volt lithium-metal-polymer batteries to be used as backup power sources in telecom cabinets, those green electronic boxes parked between suburban yards all over the country. They started shipping them in 2004, in respectable volumes—AT&T installed seventeen thousand in cable boxes used by its U-verse broadband network. But right around the same time, just as the telecom-backup batteries started shipping, Avestor laid off a quarter of its staff and pulled out of the automotive market to focus solely on telecommunications. The company's main automotive customer, a consortium of French companies that wanted to build electric cars, had bailed, saying that the Avestor batteries didn't work.

Then AT&T cable boxes equipped with Avestor batteries began exploding. The first incident, in October 2006 in Houston, blew a hole through a nearby wooden fence. The next major blast, near Milwaukee on Christmas Day in 2007, launched the enclosure's fifty-pound metal door five feet. (There were two fires as well, but the explosions naturally got the most press.) AT&T committed to replacing all seventeen thousand batteries installed in their U-verse boxes, but Avestor would be of no assistance. In a statement, AT&T said, "Normally, we would work with a vendor to diagnose problems and develop solutions. We can't do that in this case." The reason they couldn't is that Avestor had filed for bankruptcy.

The Phostech Lithium deal, meanwhile, ruined the friendship of Armand and Gauthier. Armand led the scientific team behind the company, but he quickly became disillusioned by what he saw as the unrestrained avarice spreading through the battery community. He said that Gauthier had used his laboratory to perform the work that would underpin Phostech Lithium, but without helping him to deal with a major funding crisis that occurred after 1999, and that almost resulted in his having to lay off several postdocs. But the biggest insult, Armand said, came when Phostech Lithium was unable to manufacture lithium iron phosphate powder that anyone wanted to buy. "So Phostech was created, got the exclusivity, and then they were too mediocre scientifically to do anything with the materials," Armand said. "It was of low quality, and people didn't want to use it after a few tests."

Gauthier will say little about what happened between him and Armand, citing personal, not scientific differences. As for the accusation that Phostech made inferior electrode powder, he argued that lithium iron phosphate did turn out to be "touchier" to manufacture than they had expected, so they had problems at the beginning. However, Gauthier said that those issues were sorted out, and many of the battery companies that complained about the poor quality of the company's early product buy from Phostech today. Phostech Lithium still holds the exclusive license from Hydro-Québec and the University of Texas to supply lithium iron phosphate cathode material. Everyone else is, according to Gauthier, in patent violation.

In 2004, Armand returned to France, too disillusioned to work. "I was crushed," he said. "Batteries are the only hope for changing the fuels in transport. We know that the fuel cell won't make it for years. So I mean everything is in the hands of the battery people.

"It started very idealistically," Armand continued. "People in the 1970s didn't talk about global warming, but the movement in the USA was about resources being depleted and pollution. The feeling of emergency— that I was working for my grandchildren. But now I think my daughter's going to face the problem. I've flown from Europe to California many, many times. When you fly over Greenland, one time out of every five you have a clear sky. And you used to see the glacier coming straight to the sea—and no shore, or maybe a few meters. Now you see a hundred meters, one kilometer of land between the glacier and the shore."

It was Donald Sadoway, an MIT electrochemist and a colleague of Yet-Ming Chiang, who had urged me to call Michel Armand. The afternoon after Yet-Ming Chiang's talk at the Materials Research Society in Boston, I had walked across the Harvard Bridge to visit Sadoway, and we sat and spoke for two hours in his high-ceilinged, lamp-lit office. Sadoway has the demeanor of the endearingly egomaniacal professor, the academic who became a lovable star teacher by virtue of his ability to suck the air out of the biggest lecture halls. He was wearing a black Dior jacket, a black button-down shirt, and black tie that looked vaguely like a piano keyboard. Sadoway has problems with lithium-ion batteries, primarily concerns about safety and cost. Sample comment: "Suppose I come to you and I say, 'I got this cool thing, I've got this really cool battery.' And I say, 'Well, the electrolyte's actually flammable.' And everybody's saying, 'Yeah, let's put it in a car, let's build a big one!'" We talked about his proposed alternatives, his years in the field, the story of American energy-storage research. And then, as our conversation drew to a close, he nearly jumped out of his seat and said, "You need to talk to Michel Armand! He's just a broken, broken man. The last time he was here, you know what he said? He said, 'The number-one property of lithium iron phosphate is that it is an excellent catalyst for human greed.'"

7

THE BRINK

As the Volt battery battle approached its resolution in late 2008, General Motors entered a state of terminal decline. Toyota became the biggest automaker in the world, claiming the title GM had held for seventy-seven years. Meanwhile, GM was setting records for the speed-burning of cash; the company lost $30.9 billion that year. In November, Rick Wagoner flew to Washington, D.C., on a company jet to ask for money; he came home with nothing but scars from a congressional lashing. A month later he returned to Capitol Hill, this time in a Volt prototype, and secured the emergency loan that kept the company functioning for the near future. The loan Wagoner got on his return trip to Washington was only enough to stanch the bleeding for a few months, enough time for GM to present a plan to the government that proved the company could become a viable business in the twenty-first century. Meanwhile a thirsty schadenfreude was growing among the public and the pundits, many of whom seemed eager to watch the automaker die.

Perhaps because it was the one positive public-relations chip GM had left, the Volt was kept well funded and on schedule. By then, however, it looked like a very different car. Soon after its unveiling in 2007, it became clear that the concept design wouldn't work in the real world. Publicly, most of the blame was placed on aerodynamics. As Bob Lutz put it, the show car had better aerodynamics turned around backward than it did forward, and in a car like this, "aero" would be essential for getting as much range as possible out of the battery pack. Aero is the reason for the shape of the Prius; only certain angles of a car's windshield and roofline and rear end allow air to flow smoothly across the car's body and then break away cleanly without creating vortices or low-pressure areas that produce drag. Which means that when aerodynamics comes first, a designer's aesthetic choices are restricted, a fact that explains any geometric similarities among the production Volt, the Prius, and Honda's hybrid, the Insight. The Volt spent some seven hundred hours in the wind tunnel, double what a normal car project would get, and in that time Bob Boniface and his team found dozens of tiny design tweaks that could squeeze another fraction of a fraction of a mile out of a battery charge. Guards were installed to cover the windshield wipers. Side-view mirrors got pushed away from the body of the car and placed on the end of narrow posts. A five-millimeter-high lip was added to the minimal spoiler in order to reduce aerodynamic drag by five counts and earn an extra quarter mile.

Yet Boniface says that aero wasn't the main reason for killing the concept design. The bigger issue was that to be remotely affordable, the Volt needed to share a platform with an existing car. Designing and manufacturing a new architecture in addition to all the novel engineering that was going into the Volt's power train would have added hundreds of millions of dollars to the cost of the project. The best candidate for platform sharing was the Chevy Cruze compact car, which GM planned to release in the United States around the same time as the Volt. Boniface and his designers used Photoshop to see what the car that GM unveiled in January 2007 would look like adapted for the wheelbase of the Cruze chassis. "It was ridiculous looking," he said. Instead, Boniface's studio redesigned the car, and it was unveiled at GM's one hundredth anniversary party in Detroit on September 16, 2008.

By now plenty of other carmakers were chasing the Volt's halo effect, some more seriously than others. At the Los Angeles Auto Show several weeks earlier, for example, Mitsubishi encouraged journalists to take their egg-shaped, Japan-only i MiEV electric car, which the company was talking about bringing to the United States, on a drive around downtown LA. At the same show, Mini announced that it would be leasing five hundred all-electric Mini Es, conversion cars with a backseat full of lithium-ion batteries, a top speed of ninety miles an hour, and a range of 150 miles on a charge. But the Mini E seemed hastily assembled and unserious. The handler who walked me to my test drive warned me that the strong regenerative braking was causing the car to lurch forward as soon as you removed your foot from the accelerator, and it was giving some people severe motion sickness.

Still, there was no doubt that with each successive auto show, where carmakers attempt to surprise and one-up each other with their latest and most impressive efforts, electrics, plug-in hybrids, and regular old hybrids were becoming ubiquitous.

The electric-car buzz reached a new level at the North American International Auto Show the following January. The GM and Chrysler press conferences seemed designed solely to impress any visiting Washington politicians, who had yet to announce whether they would rescue the two companies. The first day of the show, GM staged a mock political rally, hauling out hundreds of employees who carried signs saying "Here to Stay," "We're Electric," and "Charged Up" as a parade of GM's best cars and trucks rolled across the showroom floor. The next morning, Rick Wagoner made an announcement that could have seemed wonky to an outsider but was in fact the best thing GM could have done to certify the Volt's status as a real car: he revealed the winner of the battery derby.

It was a frigid, postblizzard morning in Cobo Hall. A model of the Volt's six-foot-tall T-Pack battery stood on the press conference stage, looking more than a little like a crucifix on the altar. With Lutz and the victorious battery team by his side, Wagoner declared Compact Power, Inc., the American subsidiary of LG Chem, the winner. It was the safer choice—a company that already produced millions of lithium-ion cells each month, compared to an untested start-up that hadn't yet staged a public stock offering. Wagoner delivered his remarks in the stiff, staged language typical of this kind of event, but he was emphatic about the importance of the Volt—and its successors, such as the electrified Cadillac coupe concept the company unveiled the day before—to the future of the company.

The Volt battery arrangement would be different from the usual carmaker-supplier relationship, because GM was getting back into the battery business itself. Only the Volt's cells would be made in Korea. In a new thirty-one thousand-square-foot factory that GM would build in Michigan, GM workers would assemble those cells into finished Volt packs. Going forward, battery manufacturing would become a "core competency" of the company, as essential to competing in the twenty-first century as the ability to build V-8s was in the twentieth. It was a major investment and a sober move toward commercialization, and it contrasted starkly with the fantasyland press conference Chrysler gave that week, which seemed to consist primarily of rolling existing cars onto the show floor and saying, See this? It's an *electric* Jeep.

GM's performance succeeded, at least for a day, in training the world's attention on the return of the electric car. More than that, it focused the conversation on the piece of technology that in an age of electric cars will replace the engine as most expensive, most complicated, and most essential. As a reporter for *The Washington Post* put it, "The most talked-about announcement at the North American International Auto Show yesterday wasn't about any of the gleaming cars on the convention center floor. It concerned batteries."

While the Volt program chugged ahead, General Motors arrived at the brink. The billions of dollars in federal aid that kept GM alive through the winter had come with a deadline: Show us by March 31, 2009, that you can become a viable business, or give us our money back.

The head of the White House task force responsible for assessing those plans was Steven Rattner, a Wall Street investor with no experience in the automotive industry. In October 2009, Rattner told the story of his engagement with Detroit's faltering automakers in an article for *Fortune*: "Our timing had been set by an arbitrary March 31 deadline that the Bush administration had imposed on the companies to meet a hodgepodge of conditions. And by coincidence, both companies would probably run out of money around the same time."

What they encountered when they began their urgent investigation into the state of the two troubled Detroit automakers was disturbing. "We were shocked, even beyond our low expectations, by the poor state of both GM and Chrysler," Rattner wrote. "Everyone knew Detroit's reputation for insular, slow-moving cultures. Even by that low standard, I was shocked by the stunningly poor management that we found, particularly at GM, where we encountered, among other things, perhaps the weakest finance operation any of us had ever seen in a major company." He continued:

> The cultural deficiencies were equally stunning. At GM's Renaissance Center headquarters, the top brass were sequestered on the uppermost floor, behind locked and guarded glass doors. Executives housed on that floor had elevator cards that allowed them to descend to their private garage without stopping at any of the intervening floors (no mixing with the drones). In my relatively few interactions with chairman and CEO Rick Wagoner, I found him to be likable, dedicated, and generally knowledgeable. But Rick set a tone of "friendly arrogance" that seemed to permeate the organization. Certainly Rick and his team seemed to believe that virtually all of their problems could be laid at the feet of some combination of the financial crisis, oil prices, the yen-dollar exchange rate, and the UAW.

Amazingly, when Rattner and his colleagues made their first trip to Detroit, they didn't anticipate the stir that their visit—they being the committee in charge of deciding whether two of the Big Three automakers would live or die—would generate. "Throngs of reporters awaited us at every stop while a news helicopter buzzed overhead." Rattner was particularly perplexed by the attention paid to the Volt. "More peculiarly, the ensuing press coverage seemed wildly over-focused on our test drive of the Chevy Volt, as if the company's salvation rested on this one vehicle." He went on to explain that while the team was glad to see GM pay attention to alternative technology, they were really more concerned about whether the company had a chance in hell of surviving even if the federal government invested tens or hundreds of billions of dollars. These were Wall Street guys, after all, not CARB board members. The Wall Street guys determined that the Volt and any subsequent electrified descendants "couldn't possibly have any meaningful impact on GM's finances for at least five years."

As the March 31 deadline approached, Rattner's team concluded that while GM had made significant progress in restructuring, it still had a long way to go. There was no way for them to become profitable anytime soon without the kind of changes that can rarely be made outside of bankruptcy court. A year and a half before the Volt would begin rolling off production lines, GM seemed headed for one of the largest and most complicated bankruptcies in American history.

The verdict was delivered in a public report, released at the end of March, that included a statement that no one within GM would have argued with: the Chevrolet Volt holds promise, but it's likely to be too expensive in the short term. This was something GM admitted all along: the first generation would be too expensive to make money, but that's the way it goes. You have to pay for technology on the front end if you want to lead a transformed auto industry. Predictably, the Volt's critics seized on that observation—that the first generation of Volts would lose money—as evidence that the Obama administration would kill the car.

That seemed highly unlikely. Not only had the administration made clear that it didn't want to get involved in decisions about specific products, but the current White House was more supportive of automotive electrification than any since Carter's. Obama worked an oblique mention of the Volt into his first joint address to Congress as an example of the automotive technology of the future—and as a compelling reason to fund an American lithium-ion battery industry. And on March 19, 2009, he toured the Electric Vehicle Technical Center at Southern California Edison and declared a goal of putting one million electric cars on the road by 2015. He announced a $2 billion competitive grant program for electric-car battery and component manufacturers, another $400 million for EV infrastructure, and mentioned the $7,500 tax credit already signed into law for people who buy plug-in hybrids. Finally, he delivered a pep talk. "There are days, I'm sure, when progress seems fleeting and days when it feels like you're making no progress at all," Obama said. "But often, our greatest discoveries are born not in a flash of brilliance, but in the crucible of a deliberate effort over time. And often they take something more than imagination and dedication alone; often they take an investment and a commitment from government. That's how we sent a man to the moon. That's how we were able to launch a World Wide Web. And it's how we'll help to build the clean-energy economy that's the key to our competitiveness in the twenty-first century."

Not surprisingly, Rattner and his team saw no benefit to letting Wagoner remain in charge. "It seemed completely obvious to us that any management team that had burned through $21 billion of cash in a year and another $13 billion in the first quarter of 2009 could not be allowed to continue," he wrote. On March 27, Wagoner was in Washington, and Rattner asked for a meeting. "I don't know whether Rick had any inkling of why I had wanted to see him alone," Rattner wrote. "His face was impassive as I said, 'In our last meeting, you very graciously offered to step aside if it would be helpful, and unfortunately, our conclusion is that it would be best if you did that.' I told him of our intention to make Fritz [Henderson, the vice president of GM] acting CEO and he supported that idea, cautioning me against bringing in an outsider to run the company. 'Alan Mulally called me with questions every day for two weeks after he got to Ford,' he said. As we continued our rather awkward conversation, Rick suddenly asked, 'Are you going to fire Ron Gettelfinger too?' Startled by the reference to the UAW head, I replied, 'I'm not in charge of firing Ron Gettelfinger,' and Rick soon left to brief his board on our decision." Later that day Rattner told Henderson the news. His one request: Don't do him the damage of calling him "interim" CEO. "You can fire me anytime you want, but at least give me a better chance to succeed," he said, according to Rattner.

For the Volt's devotees, particularly the thousands of people who followed every twitch of the project on gm-volt.com, an independent site run by a New York neurologist with a strong obsession with the Volt, Wagoner's dismissal raised obvious questions about the car's future. New bosses like to make big changes, right? Would the Volt fall victim to the same kind of short-term cost-cutting mentality that may have killed the EV1? Bob Lutz sent reassurances to the readers of gm-volt.com. "Thanks for your concern," Lutz wrote. "Volt will survive and prosper. We know the numbers better than the Government . . . we furnished them! First-generation technology is expensive, but you can't have a second generation without a first generation. Common sense and intelligence will prevail, here!"

To prove that the Volt was still alive, less than two months later General Motors began inviting journalists to visit Warren and drive a prototype for themselves. My visit came on a perfect spring day in mid-May, approximately twenty-four hours after the Obama administration reinforced the importance of cars like the Volt by announcing historic, strict new nationwide fuel-economy standards—a fleetwide 35.5 mpg by 2016, a 40 percent increase over existing standards.

In an immaculately clean Tech Center garage across the street from a decorative lake, Bob Lutz arrived to give some introductory remarks. "I'd just like to remind members of the media at all times, dial yourself back about twenty-seven months to the Detroit auto show of January 2007, when we showed the concept Volt and announced that we were exploring lithium-ion technology," he said. Remember the scorn, the contempt, the instant criticism from "a famous automobile company that starts with a 'T,'" he pleaded. "Here we are two and a half years later, and we are totally confident about the technology."

Volt vehicle line executive Frank Weber, a lean, towering, bespectacled German, provided further caveats. Don't pay much attention to road noise and handling, because the next year and a half is all about dialing that stuff in. Don't pay attention to the interior, because this car is, after all, far from a finished product—it's a Chevy Cruze fitted with the Volt power train. This drive was about propulsion, about batteries and motors. "The public still has the opinion that electric cars are handicapped," he said. "We want to prove that it's capable of being the first car in the household."

Our test track was GM's busy corporate campus, its lawns thick with Canada geese. Recently hatched goslings fell in step behind their mothers as they marched about the grassy lakeside lawns. The cars themselves were obviously engineering projects, with wires and cables from monitoring devices emerging here and there. With Weber sitting shotgun, I got into the driver's seat, fastened my seatbelt, and politely eased the proto-Volt around the campus as if circling a mall, looking for a parking spot. As with any electric car, the motor was silent. Any sound was road noise, a constant crunching and whining of the tires against the pavement. Weber assured me that he was not happy with the level of road noise and that much of it would be hushed out over the next year and a half.

After a few laps around a campus lake, I realized what was so remarkable about the car: that once I forgot about the novelty of the silent electric drive, and once I stopped thinking about the hype and the controversy, the Volt seemed unremarkable. In a good way. A refined, nicely equipped version of this would be a real car—not a glorified golf cart or a conventional vehicle clumsily retrofitted with a giant battery and an electric motor. The Volt prototype resembled nothing so much as a silent version of any late-model compact car.

"Go over by the lake and do some heavy acceleration," Weber said. I stopped the car at the beginning of the longest stretch of road I could find, waited until it was clear of geese, and then floored it. As we rolled forward, quickly but not wildly, the part of my brain trained to expect a crescendo of engine roar in moments like this was alarmed. Within a matter of seconds, I was at 55 mph and out of road. I did it again, psychologically prepared for the lack of internal-combustion drama, and it was exhilarating, if anticlimactic.

Back in the garage, as the small crowd dispersed, Tony Posawatz was grinning and bouncing on the balls of his feet. Like the other GM engineers in attendance, he seemed to be blocking all thoughts of financial apocalypse. Instead, he was buoyant. Next week, the first of seventy-five production-intent prototypes was set to arrive. The things the Volt team could control were, in fact, under control.

Two weeks later, General Motors filed for bankruptcy.

Across the Pacific, Nissan was preparing the first purely battery-powered mass-market car of the twenty-first century, a mysterious electric vehicle that the company promised would be cost competitive with any other conventional compact car. Nissan had been touring a battery-powered version of its Cube crossover around the United States, but like the Volt mule, the Cube was just a shell for the power train. Still, we knew what the Volt would look like; we had no idea what Nissan's electric car would look like, or what it would be called. And so in July I found myself in Japan, attending the debut of Nissan's electric car.

It had been clear for some time that Nissan was serious about its electric-car plans. Nissan and its ally Renault had announced their intention to build a purely electric car in February 2008, and since then they had been aggressively lining up partnerships with national, state, and local governments around the world in an effort to spur the construction of electric-vehicle infrastructure. By July, the Nissan-Renault alliance had made partners of governments from Denmark to Israel to Oregon to Tennessee.

Journalists from around the world descended on Nissan headquarters that weekend. The test drives happened at Nissan's Oppama Research Center, a seaside facility a half hour south of Nissan's headquarters in Yokohama, a city of more than three million people that blends into Tokyo as one vast megacity. In Oppama, the landscape was a deep wet green. The sky was pregnant with rain, and lush, green loaf-like hills stacked the coastal plain around the test track.

After coffee and a technical briefing, it was time to drive a mule version of Nissan's electric car, its drivetrain implanted in a Japanese-market Tiida hatchback. The car was powered by a floorboard full of lithium-manganese-oxide batteries, the same basic chemistry that Compact Power would be using in the Chevy Volt. The batteries were built by a joint venture of Nissan and NEC under the aegis of Automotive Energy Supply Corporation; NEC was actually the first company to commercialize lithium manganese oxide, starting in 1995.

As I sat down in the driver's seat, my Japanese handler, sitting in the passenger seat, seemed nervous, as if he half suspected that one of us gaijin were likely to drive off the track and start yard farming. I pointed the car through a matrix of traffic cones, eased onto the track, and hit the electron-dispensing pedal. Acceleration was brisk, and before I knew it my passenger was asking me to slow down. On this slick professional track, the silent ride gave the sensation of gliding. The experience was much like driving the Volt, but with far more interesting scenery. As we circled the track, a brown, bearded-looking raptor glided lazily overhead.

After a bus ride back to Yokohama, we visited Nissan's new global headquarters, an airy, ultramodern building that gives the impression of being 99 percent glass. The official christening of the building was scheduled for the next day, and the construction crews were still working to make the place presentable. Construction workers' cigarette smoke hung in the air, and the smoky-café smell was wildly incongruous in this soaring, open, ecofriendly office building.

We were ushered into an auditorium, and there it was on the stage: a bubbly, cornflower-blue hatchback that we weren't allowed to photograph and whose name we weren't yet allowed to know. The car was cute but covered in odd design flourishes, like bulging-eyeball headlamps that dominated the front end, and an oddly concave rear. The interior was attractive and subtly high-tech, with digital gauges, touch-screen navigation, and a glowing blue half sphere for a gear shifter that looked as if it were designed to read the driver's palm and power the car on his or her karma. The car's GPS system would track its state of charge and communicate with a central data center, which would allow it to display your range at all times on the navigation screen as a highlighted radius around your current location.

The official reveal of the car happened the next day, a Sunday, during the inauguration of the company's new headquarters. On the ground floor of the soaring-ceilinged atrium, Nissan staged an auto-show-style press conference, in which blandly pleasing finger-picked guitar music played over the loudspeaker as secret doors in the light-studded wall behind the stage parted, and Nissan's electric car—the Leaf, we then knew it was called—emerged.

The Nissan Leaf was even riskier than the Volt. To be fair, Nissan wasn't in government receivership, but the Leaf was an enormously expensive bet that the conventional wisdom about electric cars—that they'll never catch on because people won't buy a car that has limited range and takes hours to repower once its electron tank is empty—was simply wrong. Nissan had taken a holistic approach to the electric car, recruiting the kind of allies that could build the necessary infrastructure. The company also made stunning claims about the car's cost—that it would be priced about like any other car in its segment, and even at that price, it would be profitable in its first generation.

CEO Carlos Ghosn and a cast of Japanese eminences, including the former prime minister of Japan, Junichiro Koizumi, and the mayor of Yokohama, sat inside. Ghosn gave his speech in English. "The inauguration of Nissan's global headquarters and the arrival of the Nissan Leaf are exciting events in the life of our company," he said. "Both are clear signals that Nissan is decisively turned toward the future." He spoke of a "zero-emission future," a "new era in the automotive industry." The Leaf, he said, was the first step.

"From the outside this family hatchback may appear as another real-world attractive Nissan car, born and developed here in Kanagawa Prefecture. But in fact this car represents a real breakthrough. For the first time in our industry history a car manufacturer will mass-market a zero-emission car, the ultimate solution for sustainable mobility. As its name suggests, the Leaf is totally neutral to the environment. There is no exhaust pipe, no gasoline-burning engine. There is only the quiet, efficient power provided by our own compact lithium-ion battery pack."

By the summer of 2009, the idea that the Volt was nothing but a concept had been replaced by broadsides against what the production vehicle had become—its looks, which plenty of people thought were bland and pedestrian compared to the 2007 show car; the very concept behind its drivetrain; and so on. The bizarre ongoing debate over the Volt, unprecedented for a car that was still a year and a half away from production, recalled a comment that George Westinghouse made during the late-nineteenth-century "War of the Currents": "Thousands of persons have large pecuniary interests at stake, and, as might be expected, many of them view this great subject solely from the standpoint of self-interest."

Each automaker held differing yet fairly predictable views on why the Volt made no sense. The critiques were predictable because they always favored the technology in which the carmaker had invested the most money. Although Toyota had announced plans to build a test fleet of five hundred lithium-ion-powered plug-in Priuses, in May Bill Reinert of Toyota told a National Academy of Sciences panel that plug-ins were still hamstrung by high cost and battery issues. Carmakers such as Mercedes-Benz and Audi, on the other hand, who had been working for years to perfect clean diesel engines that could meet American particulate-pollution standards, tended to argue that while electric drive was interesting, diesel was the best way to quickly reduce petroleum usage. In September 2009, Johan de Nysschen, president of Audi of America, called the Volt a car for "idiots." Smacking electric drive in favor of the diesel engine sometimes drove the employees of the big German automakers to an irrational, almost faith-based dismissal of electrification. I once talked to an engineer for Mercedes-Benz who was adamant that all batteries, no matter what they were made of, were environmentally disastrous. What are we going to do with all these batteries? Dump them all in landfills? He asked me the question with a smug, almost contemptuous grin. Well, they can be reused, or recycled. But they're toxic! he said. Not if they don't contain any toxic metals, I replied. I don't care! They're batteries!

The highest-profile spat began when Elon Musk appeared on *The Late Show with David Letterman* on April 29, 2009. A hoarse Letterman grunted out a fawning introduction. "This is gonna be exciting, because here's a guy doing something to try and make things a little better for the eight billion humans who inhabit this planet," he said. "He has been called the Henry Ford of his generation. He is the CEO and chairman and product architect for Tesla Motors, currently producing the only highway-capable electric vehicle available in North America. Please say hello to Elon Musk."

First, Letterman and Musk ran through the basics. The cost of the car: $100,000. The reason for starting with a high-end sports car: "The goal has always been to make low-cost cars, but when you have new technology it takes time to make it lower cost and mass market," Musk said. "If you think of the early days of cell phones or laptops or any new technology, it starts off expensive. You remember that giant phone that the guy in *Wall Street* walked down the beach with, and that was cutting-edge technology—well, it's the same thing with cars."

Letterman pointed out that electric cars are not exactly a new technology, and Musk delivered a one-sentence version of the history of the competition between gas and electric. The reason gas won, Musk said, was the range issue. "Now with the advent of lithium-ion batteries, we can now address the range issue."

"So the real breakthrough is the batteries, then," Letterman said. "How to store that electricity."

"Yeah, it's the single biggest breakthrough."

Letterman brought up the EV1. Musk replied by referring to *Who Killed the Electric Car?* "Chris [Paine, the film's director] shows how much people really wanted EV1," Musk said. "They wanted it so much that when the cars were forcibly taken away from them and crushed, people who had those cars held a candlelit vigil for the destruction of those cars. Now, when is the last time you heard of anyone doing a candlelit vigil for the destruction of any product, let alone a General Motors product?"

This line drew a sustained round of laughter. The trash talk quickly escalated, and almost all of it came from Letterman.

"What if the electric car movement had not been killed off twenty years ago? Would that be a means of keeping these factories"—GM factories—"open today?"

"With the benefit of hindsight, General Motors probably wishes they had done an EV2 and an EV3 after the EV1 instead of crushing them," Musk said.

"Doesn't it make you a little angry that we're all pretending like, 'Whoa, we got an electric car!'" Letterman said. "It's not like someone landed from Mars. We had them at the turn of the century! It's frustrating to me. Is it frustrating to you?"

Yes, Musk said, it is. But he made a gesture of goodwill toward GM. "I really thought that the incumbent car companies would do this," he said. "Bob Lutz of General Motors was kind enough to credit Tesla with the inspiration for the Volt. You know they're coming out with a plug-in hybrid, the Volt."

And the Volt insults began. "The Volt has a range of forty miles," Letterman said. "That'll get you down the driveway and back. 'I gotta go pick up— Take the electric car! Call me if there's trouble at the curb!' I mean it's insane! I mean forty miles is the range on the Volt, that's ridiculous, isn't it ridiculous? And General Motors is all, 'Oh, boy, we got the electric car.' I mean, that's crap!'"

Wild applause.

Commercial break.

"I drove your car," Letterman said after returning from commercials. "I was skeptical. I thought, 'This is for guys in Topanga Canyon selling sprouts.' But the thing is bulletproof. I thought that the first time I charged the thing my house would catch fire. It goes like a bat out of hell," he said. "Between you and me, the first time I drove it I was worried that it would magnetize my nuts."

Musk laughed awkwardly.

"You know, fuel-cell cars, that's a load of crap too," Letterman continued. "You know: Oh, hydrogen! It's gonna make its own hydrogen! They're talking about twenty years from now. *Maybe* twenty years from now. So I think that these automobile companies—and it couldn't be a worse time for them to be doing this—are just shining people on. Because if they were actually working on technology that was gonna be in showrooms, they wouldn't have to be closing down plants and filing for bankruptcy."

Once again, wild applause.

And in Detroit, wild fury. Mark Phelan of the *Detroit Free Press* published ten things Letterman should know about the Volt. About a month later, GM got Bob Lutz on *Letterman* to run defense. Letterman began by asking about the EV1—more politely, now that he was in the presence of a company man. Respectfully, if the EV1 had continued, wouldn't that have kept the company's financial problems away?

"I'd like to say yes, but the answer is really no," Lutz said. The batteries weren't ready, the cars cost $100,000 apiece to build, the maintenance of the fleet was costing a fortune, and the finance guys eventually said, 'Enough.'" He continued, "What has only recently become available is a battery that will store enough energy" to make an electrified car practical.

"The problem has been the availability of lithium-ion batteries," Lutz continued. "The Tesla uses 6,831 laptop batteries all wired together." The Volt, by contrast, uses a specially designed automotive cell, and that's why it will sell for around $40,000 before tax credits,

Lutz explained. Then Lutz and Letterman made nice as Letterman asked to buy the first Volt. "I've probably promised it to about seven or eight other people," Lutz said.

Next, Letterman and Lutz walked over to a side stage and the production Volt rolled out. The audience applauded enthusiastically, but there were far fewer ooohs and aaahs than when Tesla's prototype Model S sedan appeared on the show.

And then Letterman concluded the segment by reaching inside the car, grabbing the steering wheel, and shrieking, "Ahhhh! Ahhh! I'm being electrocuted!"

New Year's Day 2010 marked the beginning of the home stretch for the Volt. Eleven months and the first Volt had to roll off of a production line. It would be a soft launch; the first cars would go to utilities and governments and various test fleets. It would initially be available only in California and a few other markets. It still wasn't clear whether the car would be sold or, like the EV1, just leased. In fact, GM's talk of test markets and "learning how consumers use the car" was uncomfortably close to the language used to launch the EV1.

The newest set of Volt prototypes, approximately eighty of them, were being flogged at test tracks across the country, piling up miles, the engineers' fingers crossed that no showstoppers emerged. So far, though, the Volt team had been nothing but pleased with the results of their torture testing. To check the car's hill-climbing ability, Volt engineers drove a prototype to the top of Pikes Peak and back down. They tested the battery's liquid cooling and heating system in Death Valley and Kapuskasing, Ontario, where, according to Tony Posawatz, the engineers are disappointed if the temperature climbs above −18°C. Not that they would be likely to admit any major mishaps, but the Volt principals appeared convincingly confident. What was left for the final eleven months, in addition to scaling up the manufacturing apparatus needed to put the car into production, was software work—battery management, human-machine interface tweaks, the final debugging and software coding.

By the start of the North American International Auto Show in

January 2010, the Detroit-area plants that would assemble the Volt and its battery packs were warming up. Soon, both would begin their rehearsal run, building batteries and cars for testing and validation. On the first day of the show, I asked Jon Lauckner what he thought of the electrification theme that had overtaken Cobo Hall. "It's interesting and gratifying," he said. He thought back on the announcement of the Volt in this same building three years earlier. "At the time, people were enthusiastic about the fact that we had reclaimed the electric car from the ash heap of history, but there were quite a few skeptics out there who said, 'Nah, we don't think they'll go to production with this vehicle,' because, in their words, batteries weren't included. It's gratifying, to say the least, that something that started with a concept car in January of 2007 has now spawned many, many variants of the same theme— that is, drive using electricity rather than petroleum."

The intervening months hadn't been without drama. In December, Frank Weber left the Volt project to return to Opel in Germany. Most significantly, Fritz Henderson was out, the second CEO firing that year. The timing was atrocious. The first week of December, Henderson was scheduled to give the keynote speech on the opening morning of the Los Angeles Auto Show. He was canned the night before. Chairman Ed Whitacre took his place as interim CEO, and Bob Lutz replaced him as the morning speaker.

In LA, I caught up with Posawatz and asked him what he thought about the latest shake-up. "Surprised," he said numbly. "A little disappointed. Don't understand the timing. But there's been all kinds of shake-ups this last year. I mean, literally a year ago at the LA auto show I can remember watching on TV our former CEO Rick Wagoner getting grilled by the congressional panel." Posawatz said he was not at all concerned about the latest major personnel change harming the Volt program. The Volt had too much momentum. It was too close to the deadline. "The Volt team will outpace much if not most of any General Motors restructuring," he said. "Remember: By next month I'm building batteries. By the early spring I'm building cars in Hamtramck. We have a demonstration fleet program which was announced earlier today. We're gonna do stuff in 2010, and then when we're ready to produce the production car we also peel off some for customers, some for the DOE. We'll collect data using OnStar, which no one else can do. This machine is in motion. It is humming at a fast pace. And oh, by the way, having been involved with a few short discussions with Mr. Whitacre himself—huge fan of the Volt," Posawatz said of the chairman and new interim CEO, who had given himself the job when he fired Fritz Henderson. "*Huge* fan of the Volt."

Not to say Posawatz showed no signs of burnout. He took a little time to vent about the approach that many journalists were taking to the Volt. "I'll be honest with you, I am personally very disappointed in your craft," he said, "that the most oft-asked question on this program since day one is: 'What will it cost?' And today I told these guys: You guys are being lazy journalists. Here we had rides of the vehicle, we showed off the extended-range mode, et cetera et cetera. We're building a battery plant that will start building batteries next month, General Motors, U.S. green jobs, and you guys are asking the lazy question. Why don't you guys tell me? Where is California with their HOV lane legislation? What does a Prius with an HOV lane sticker sell for versus one that doesn't have one?"

In their defense, I said, the cost of the car is a major issue here, because if you can't sell it for less than $40,000—

"Uh-uh, they ask about *price*, not cost," he interjected. "I'd love to talk with them about cost." And so we talked cost, and I asked him to respond to the common claim that lithium-ion batteries will never be cheap enough to make electric cars a mass-market reality. "It's not a Moore's law curve on batteries. It is not." Nonetheless, Posawatz added, evolutionary engineering changes will soon make the batteries vastly cheaper. Set fundamental science aside for a moment. The Nissan Leaf's battery is air-cooled, which is cheaper than the elaborate liquid conditioning system the Volt uses; maybe they'll find that they can use air to cool the Volt's battery pack? The Volt is designed to use only half of its battery capacity, a conservative decision meant to ensure that the batteries can last the full life of the car's warranty. But in time it may become clear that you don't need to set aside a full half of the battery as cushioning. If you can use more of the same battery, you just increased the range of the Volt without spending anything. "Can I take thousands of dollars out of the battery in a couple years?" Posawatz said. "You bet. *Thousands* of dollars. Am I ahead of all the experts' opinions relative to the dollars-per-kilowatt-hour cost? I'm ahead of all the experts—who haven't built a battery in their life. Right?"

In Detroit the following month, as Lauckner and I talked, a mob of people crushed through the Volt stand—the kind of human swarm that at the big auto shows generally signals the arrival of a celebrity. In this case it was a congressional delegation, led by Nancy Pelosi, there to see how the government's investments were performing. GM's latest CEO, Ed Whitacre, escorted Pelosi, John Dingell, and Steny Hoyer toward the Volt, and they were swarmed by video cameras and photographers.

Beyond the unusual level of government interest, plenty of other things at this year's Detroit show emphasized that the landscape was shifting. On the other side of a wall from the GM stand was an unusual competitor: the Chinese company BYD, which began its existence in 1995 as a battery manufacturer and had only recently moved into automobiles. From their base in the industrial megalopolis of Shenzhen, an hour outside Hong Kong, BYD was plotting domination of the world electric car market. They had allies in Warren Buffett (who in 2008 became a 10 percent investor in the company) and the Chinese government, which in April of 2009 declared a national goal of becoming the world's largest supplier of electric cars by 2012.

What BYD had in high-profile backing it lacked in polish. By the glossy standards of the international auto shows, its booth was chintzy, with ragged carpet edges and a ring of tiny plastic potted plants surrounding its marquee product, the e6 battery-powered sedan, which it said it would soon bring to the United States.

Even the Germans were caving in to electrification. That day BMW revealed an electric version of its 1-series coupe, a bridge between the Mini E project (which had been plagued by poor battery life in cold weather, long waits for customers to get their garages wired for charging docks, and other problems) and a high-end electric urban commuter that BMW planned to release by 2015, which they were calling the MegaCity. Across the aisle, Audi was showing off a shortened coupe version of the E-tron concept, the second electric supercar it had trotted out in the past two months.

Tesla stood across the aisle from Mercedes. A dusty white roadster with several Sharpied autographs testified to Tesla's position that the charging-infrastructure issue was overblown, and that range anxiety was "for the weak," as the company's press materials put it. Tesla engineers had driven the roadster from Los Angeles to Detroit for the show, and they intentionally left it unwashed. The following day Elon Musk gave a press conference that was almost affected in its contrition, that seemed designed to begin correcting the image he had built for himself since taking over as Tesla CEO as a brash, trash-talking egomaniac who loved being quoted calling his critics "douches." While the new plant that would build the Model S sedan was still delayed, Tesla seemed to have righted itself. It had sold more than one thousand roadsters since 2008 and—no small feat—it had survived the most devastating recession since the Great Depression.

At the end of the first day, the visiting congressional delegation took to a small stage on the showroom floor to give a few remarks. Nancy Pelosi, Steny Hoyer, John Dingell, Ray LaHood, Hilda Solis, and a few other Washingtonians had been chauffeured throughout the day by Michigan's governor, Jennifer Granholm. The succession of stiff, brief speeches consisted of a series of platitudes about the Survival of an Industry and Keeping America Number One. The most sincere-sounding moment came when Steny Hoyer, House majority leader and representative from Maryland, admitted to feeling like a "kid in a candy shop" on the show floor, wistfully recalling that when this elder congressman was a kid, the "symbol of America's greatness was the automobile."

8

THE STIMULUS

As the world's automakers began to cautiously embrace the lithium-ion-powered electrified car, the battery boom began. In 2008 and 2009, think tanks and industry groups issued forecast after forecast assessing the electric car's chances this time around, and, of course, their conclusions differed wildly, ranging from profoundly pessimistic to delusionally optimistic. But in the battery business everyone knew that even a small electric-car market would be a tremendous opportunity. Even early-adopter volumes of lithium-ion-powered cars of any kind—plug-in hybrids that run on gas after ten or twenty or forty miles, pure EVs, next-generation variations on the basic Prius format—would require a staggeringly large number of batteries. Charles Gassenheimer, the CEO of Ener1, parent company of the American lithium-ion manufacturer Enerdel, liked to say that one hundred thousand electrified cars a year—just twice the number of Leafs Nissan will build for 2011—would "soak up the entire cell capacity that exists in the world today for the cells you have in your laptop computer."

Advanced-battery start-ups were popping up like mushrooms after a spring rain. Even large automotive suppliers with a steady stream of business in other areas wanted in. Johnson Controls, for instance, an automotive supplier and manufacturing conglomerate with three hundred thousand employees worldwide, decided it needed to enter this new market, and so it sought out a partnership with the French battery company Saft. The Canadian auto supplier Magna began marketing itself as an electric-car-battery packager. Dow Chemical got into the field by forming a joint venture with Kokam, a Korean battery company.

Soon the word "battery" was no longer exclusively associated with the Energizer Bunny and the TV remote. Batteries became a segment of the clean-tech boom, with all the dewy and righteous credibility of thin-film solar and offshore windmills. If anything, the battery industry carried an additional dose of sexiness because of its attachment to futuristic cars like the Tesla Roadster.

National, state, and municipal governments around the world became interested in electrification. Politicians from Israel to Oregon shook hands with executives from the Renault-Nissan alliance, joining in an effort to build charging stations and other EV infrastructure. Austin and Indianapolis began competing with Detroit to become advanced-battery hubs, players in the next big technological thing. Better Place, a battery-swapping and EV-infrastructure company founded by the Silicon Valley entrepreneur Shai Agassi, was circling the globe negotiating electric-car infrastructure deals.

American businessmen and politicians learned to work the battery issue with a simple, persuasive line: "We're going to end up replacing foreign oil with foreign batteries!" The American auto industry had been kneecapped by the global recession and its own structural imbalances and mismanagement; the lean Detroit that it would take to survive could never make up for all the jobs that had been lost by building cars alone, and so the notion of building a sister enterprise—an American advanced-battery industry—was irresistible. Detroit wouldn't just build cars; it would build a new breed of car *and* the batteries that go into those cars, along with the solar panels and windmills that will charge those batteries.

By the end of 2008, the heads of Detroit's automakers were agitating for government help to reinvent the Rust Belt. William Clay Ford, Jr., recited the standard national security line to *The New York Times*: "We really don't want to trade one foreign dependency, oil, for another foreign dependency, batteries." Rick Wagoner, then still at the helm of GM, made much the same argument to *The Washington Post*. Charles Gassenheimer spoke on behalf of the nascent North American battery industry. "Advanced batteries are as important to this new market as the microprocessor was to the emergence of the personal computer revolution," Gassenheimer told reporters. "You can't have one without the other. Unfortunately, the U.S. manufacturing capacity is just not there today to produce them in anything approaching the numbers we need over the next few years at costs the market demands."

It didn't take long for this talk to translate into government support. James Greenberger, the director of the National Alliance for Advanced Technology Batteries, a not-for-profit trade organization, recalled a conference he held in Chicago in June 2008. "We invited industry and local political leaders from the Chicago area," Greenberger said. "The takeaway was that whereas lithium-ion technology was a potentially transformative technology and might, in fact, become the basis for the auto industry and permit us to displace a lot of imported Middle Eastern oil, we were likely to displace that Middle Eastern oil with Asian batteries." He says he spoke to a fellow "who worked for one of our local politicians" about how best to build this new industry. Have another conference? Start an industry group? Eventually, with the politician's encouragement, Greenberger started assembling NAATBatt, as it came to be called.

"Not much happened in the balance of 2008," Greenberger said. "But then this fellow's boss got elected president in November 2008, and all of the sudden everybody started paying attention."

The Obama administration was the best friend the battery industry could have wanted. In February 2009, during his first address to a joint session of Congress, President Obama cited the insult of running American-built plug-in hybrids on "batteries made in Korea" as evidence that the United States was falling behind in the clean-energy race. Obama had already set an official goal of putting one million plug-in hybrids on the road by 2015. The opportunity to do something big in that direction came in the form of crisis. As the world economy teetered on the verge of collapse, the Obama administration saw the opportunity to use a necessary package of emergency spending to build a lithium-ion industry—a factory for clean energy jobs—from scratch.

"It must have been January or February, I got a call from Rahm Emanuel's office, saying, 'We've got this stimulus package coming down, and we've decided we want to do something for battery funding in the stimulus package. How much do you guys need?' And I thought, 'Wow, that's an interesting question.'" After talking to a few colleagues, Greenberger knew the answer. "We'll take $2 billion," he told Emanuel's office. "It was a little bit tongue-in-cheek, but that was certainly the wish. And lo and behold, that's what ends up in the stimulus package."

In August, the Department of Energy revealed the winners of $2.4 billion in stimulus funding, and Obama's deputies were dispatched nationwide to deliver the news. Secretary of Energy Steven Chu was sent to Celgard, a Charlotte, North Carolina, company that makes the thin polymer separators that lie between the electrodes of a lithium-ion battery. "This is the single largest investment ever made in advanced battery technology anywhere in the world," Chu said of the grant pool. Awards like the $49 million that Celgard was receiving were "incredibly effective investments that will come back to us many times over—by creating jobs, reducing our dependence on foreign oil, cleaning up the air we breathe, and combating climate change. They will help achieve the President's goal of putting one million plug-in hybrid vehicles on the road by 2015. And, most importantly, they will launch an advanced battery industry in America and make our auto industry cleaner and more competitive."

If the stimulus money didn't pick winners, it at least narrowed the field. Dozens of companies were missing from the list. The four biggest awardees all had plans to build factories in the Detroit area—Johnson Controls, A123, Dow Kokam, and Compact Power, the operation that was supplying the batteries for the Chevy Volt. In a close fifth place was Enerdel, which already had a functioning cell factory in the motor-friendly city of Indianapolis.

Plenty of people saw politics at work in the list of winners. "I think, in the end, we were perhaps not in the state that got rewarded," said Christina Lampe-Onnerud, CEO of the Massachusetts lithium-ion manufacturer Boston-Power. "We didn't have a lot of political insight on that. We came with more of a technical opportunity to produce something really good—but I totally respect and understand that's a different process." She said her company had been planning a new factory in China, but that when they applied for stimulus funding they put those plans on hold. "The day we were informed of [the stimulus decision], we continued with our China plans. So those three thousand jobs are going to China."

On a rainy, near-freezing day in January 2010, dozens of reporters gathered on the polished concrete floor of Enerdel's lithium-ion cell factory, which at the time was the only remotely commercial-scale lithium-ion cell plant in America. A wall of safety-goggled Enerdel workers formed a semicircle behind us. It was like a scene out of an inspirational commercial about American Ingenuity and Green Jobs. Charles Gassenheimer strode onto a small stage and declared that the Indianapolis area would soon be known as the "Silicon Valley of the automotive industry." He then introduced Mitch Daniels, the governor of Indiana, a wiry man in a chipper mood.

Daniels took the microphone. "I've always been an internal combustion guy," he said. "But I have to tell you, I've become converted in every way to be an enthusiast about electric vehicles—what they can mean to our nation's future, the world's environment and economy, but, specifically, the economy of our state."

Enerdel is the automotive-grade battery arm of the New York–based company Ener1. It was formed in 2004 when Ener1 purchased the lithium-ion battery assets of the auto supplier Delphi, including this factory, which, incidentally, Delphi used to develop parts for GM's EV1. Five months before this presentation, Enerdel had been granted a $118.5 million chunk of stimulus money. Today, Enerdel was announcing that it was using its share of the pie to build a third Indianapolis facility, a much larger factory that by 2015 could churn out enough lithium-ion batteries to power 120,000 electric cars a year.

After his pep talk, Governor Daniels yielded question time to Gassenheimer, who used the first reasonable opportunity to frame Enerdel's mission in the language of national security: "We shouldn't trade our dependence on foreign oil for foreign battery manufacture." As if reviewing the benefits of a battlefield airstrip, Gassenheimer explained Enerdel's logistical superiority. Shipping several-hundred-pound battery packs that are considered class-nine dangerous cargo is going to be expensive, and with factories in Indianapolis, Enerdel is convenient to any automaker that might conceivably want to build electrified cars in the United States—both Detroit's Big Three five hours to the north and the Asian and European automaker factories in the South.

Short, bespectacled, and perpetually wearing a dark suit, Gassenheimer has the looks and manner of a New York hedge fund guy. Which is appropriate, because that's what he is. Before he joined Ener1 in 2006, Gassenheimer was an asset manager, and in that role he financed Ener1's purchase of Delphi's lithium-ion battery business. For the first two years, the new venture blew through money. It had an "upside-down balance sheet," Gassenheimer said. In January 2006, the board of directors asked him to come on as CEO, and he made the leap into his current role as a full-time evangelist for the energy storage business.

We had had lunch in Manhattan a few months before the event in Indianapolis. He chose the Penn Club, where he is a member, and sitting in the club's high-ceilinged, wood-paneled dining room he explained to me how the stature of the lithium-ion battery business had grown over the past three years.

"You can track it this way," he said. "In 2007, I'd go to Washington and I'd get a bunch of meetings with a bunch of junior energy-policy advisers of various congressmen and senators. And in 2008, it was mostly folks who had to take the meetings—they didn't want to take the meetings. The Indiana delegation, the true believers"—for example, Congressman Jay Inslee of Washington state's first congressional district. "By the end of 2008, this was starting to get a lot of attention—big dollars, a $25 billion direct loan program [under the Advanced Technology Vehicles Manufacturing Incentive Program]. All of the sudden it was like, $25 billion for green cars, holy shit, we better find out what's going on. And this year I've met with everybody but the president himself. It's achieved that sort of rock star status of, they're calling you: 'Come to Washington, please come meet with Browner, please come meet with Rahm.'"

Now, the following January in Indianapolis, I was getting a close look at the business behind the hype. Earlier that morning we had toured Enerdel's existing cell-manufacturing line, which would soon be expanded and then replicated in the new facility on a much larger scale.

Even in this all-American factory, the process began with electrode powder imported from China. Electrode powder contains the active raw materials that make a battery run—industrialized, proprietarily tweaked variations on the basic lithium-based compounds that the John Goodenoughs of the world cooked up in their laboratories decades earlier. At Enerdel, in the first step of battery-cell manufacture, a clear liquid solvent is piped into a mixing room, where it's combined with electrode powder, carbon black, and a chemical binder, and then blended in giant gleaming metal mixers that look like industrial-strength pizza-dough machines.

Next, a sort of industrial printing press paints the electrode slurry onto long sheets of metal foil—aluminum for the positive electrode and copper for the negative. The freshly painted sheets of electrode material then pass through an oven for curing. Because Enerdel makes prismatic, or rectangular batteries (as opposed to "wound" cylindrical cells), next a set of machines chops the reels of electrode material into book-size rectangles. To ensure that no gremlin particles are hiding on the surface of the electrode, the sheets are then brushed and vacuumed.

In a climate-controlled dry room, positive and negative electrodes get sandwiched together with a "separator" in between. The separator, which looks like nothing more than a strip of white trash-bag plastic, is forgettable but essential: it keeps the two electrodes from touching, preventing a short circuit while allowing them to pass ions back and forth. These electrode-separator sandwiches are next dropped into plastic pouches, which are then filled with liquid electrolyte and vacuum-sealed shut. The result is a cell, the building block of a battery.

In another room, the cells are placed in what look like plastic milk crates and inserted into chest-high metal vaults. Here's where the electricity comes in. First the cells are precharged—given just enough juice to start the chemical reaction. Then the pouch is opened to release any gas that has formed; after that, it's vacuumed and resealed. Next the cells are charged to 60 percent capacity and aged for fourteen days, to make sure none of them are faulty.

The final step is a matter of wiring any number of cells together into a finished battery pack. With the help of assembly machines, workers add the cooling and heating mechanisms, the voltage-monitoring circuitry, the thermistors that detect any individual overheating cells. Cells are bundled together to form modules; modules are then bundled together in a case and wired with additional monitoring circuitry. The finished product is the black box of electricity that powers the car.

From Indianapolis I flew to Boston to visit Yet-Ming Chiang and A123. He picked me up at my hotel on the MIT campus in his Prius, which had been converted into a plug-in hybrid powered by a lithium-ion battery pack.

Chiang is an energetic man who enjoys showing off his toys, and on that day we started with his lab at MIT. We parked downstairs in the notorious Frank Gehry–designed Stata Center, an architectural landmark whose roof reportedly leaks water like a tin shanty, and took the elevator to Chiang's office, where he opened a cabinet of curiosities and proceeded to hand me one mystifying object after another, each one an example of the power of materials science to warp matter through the careful tweaking of its structure and composition.

Chiang pulled out a ghostly, Starburst-size Jell-O cube of slate-blue substance that looked like frozen smoke—a solid that looked like a puff of gas. This material, called aerogel, is used as insulation in aerospace applications. Next he handed me a football-size chunk of iridescent sodium borosilicate glass, which is used in the mirrors of the largest astronomical telescopes on the planet. Then he held up a small plastic case containing an ivory-hued carving of the Taj Mahal approximately the size of a mouse. It was the work of a 3-D printer, which translates digital instructions into a physical plastic model.

We walked across the hall to his lab, a greasy chemical shop equipped with argon-filled glove boxes and furnaces that run as hot as 2,000°C, which Chiang and his students use to synthesize materials. Chiang says he has more high-temperature furnaces than anyone else at MIT, and probably more than any lab in the world.

We hurried down the stairs to Chiang's ecocar, and as we drove away from the campus, Chiang gave me the brief version of his biography. He is an MIT lifer. He was born in Taiwan in 1958 but raised in New Jersey, Connecticut, and New York, and in 1976, following after his older sister, he began his undergraduate education at MIT, majoring in materials science and engineering. One of his first jobs was sifting fly ash from a coal-fired power plant; the fly ash was to be used in a blue-sky energy-crisis-era project involving magnetohydrodynamic reactors, which shoot high-temperature plasma through a magnetic field to create electrical current. After graduation, he enrolled in a Ph.D. program at MIT, which led directly to a job here as an assistant professor. Aside from a few sabbaticals, he never left academia for industry, partly because by the time he finished grad school, in 1985, large corporations were cutting back on basic research. "When I was a grad student, nobody thought in terms of starting companies," he told me as we drove along the Charles River toward A123's headquarters in nearby Watertown. "You couldn't conceive of taking on the IBMs of the world." Then came the Bayh-Dole Act, which allowed universities to retain title to inventions that were the product of federally funded research, and to license them to, say, entrepreneurial-minded professors. One of Chiang's first start-up ventures was American Superconductor, which he cofounded in 1987 with three other MIT faculty members. American Superconductor, which is still in business, was an attempt to use the high-temperature superconductors that in the late 1980s had snared the minds of the world's materials scientists to build commercial-scale superconducting wires, motors, and other applications.

Even before the 2002 publication of his controversial *Nature* paper on lithium iron phosphate, Chiang and his partners had begun spinning that research out into a company. They started with private capital and the $100,000 small-business grant from the Department of Energy, soon won a $750,000 DOE grant and the contract with Black & Decker to supply power-tool batteries, and after that the company grew quickly. Next came two contracts totaling $15 million with the United States Advanced Battery Consortium, an electric-car battery alliance among the Big Three automakers and the DOE that dates back to the electric-car boomlet of the 1990s. Before long they had purchased the Canadian company Hymotion, which sold battery packs for converting conventional hybrids into plug-ins, and built the lithium-ion-powered Killacycle, the quickest electric motorcycle in the world and an excellent PR device.

A123's loss to LG Chem in the Chevy Volt battery contest didn't do any noticeable damage to the company's spectacular initial public offering, which after a long delay was held on September 24, 2009. The IPO was surrounded by hype; *Mad Money*'s Jim Cramer said it could be the "hottest IPO of 2009." The automotive lithium-ion market, Cramer explained, had the potential to grow from some $31.9 million today to $74 *billion* by 2020. "That's the incredible growth hedge funds and mutual funds crave, and they're going to play it with A123," read a quotation from Cramer on his website. Skeptics pointed out that since its founding, A123 had lost $146 million on earnings of $168 million. But that didn't matter the day A123 went public. The company began by offering shares at $13.50; by the end of the day that price was up to $20.29, and A123 had raised $378 million. The IPO begat an endless series of bad electrical puns—A123 is "charged up," delivering a "jolt" or a "shock" to the markets—and brought the company esteemed investors such as General Electric, Qualcomm, and Motorola.

A few months after my visit, A123 announced plans to move its headquarters to another Boston suburb in 2011, but at the time, A123 central was located in a nineteenth-century cannon factory converted into a high-end office and shopping center called Arsenal on the Charles. It was postindustrial chic compared to the drab Midwestern manufacturing vibe of Enerdel. The executive offices were bright and airy, with young, casually dressed employees and a well-stocked vending machine where everything costs a quarter. "It used to be free, but we find that by charging just a little, people waste less," Chiang said. Bronze patent plaques and framed photos of company execs with notables such as Jay Leno, George W. Bush, and Rob Lowe lined the wall. A recent addition to the collection showed company president David Vieau at the NASDAQ listing ceremony in Times Square.

Chiang introduced me to nearly everyone we ran into, including the company counsel, Eric Pyenson, who obviously had no idea that Chiang had agreed to give me a full-day tour, unmediated by PR people or lawyers. "Is this for a piece?" he asked as he shook my hand and gave me a wary look. He quickly launched into ground rules and warnings about what Chiang was allowed to say and told me that he needed to be able to review quotes, to which I said sorry, but no. "Well, then, you're just going to have to be very aware, very conscientious of what Yet is allowed to say, otherwise I wouldn't allow him to talk to you at all."

Chiang turned to me and grinned. "Sorry, this is actually not why I brought you by here."

On Pyenson's wall were autographed photographs of presidents Reagan, Bush the Elder, and Clinton. Before law school, he worked for the U.S. government. Counterterrorism. Telling stories about his previous professional life seemed to relax Pyenson, who was soon talking excitedly about the prospects for A123 and the lithium-ion business in general. "It reminds me of the dot-com start-ups," he said. "Now it's a similar situation."

We left his office and walked across the parking lot to the research-and-development plant, which occupied a renovated power plant. With its tall brick smokestack and full wall of windows, the building looked more like a high-end Boston brewpub than a battery factory. The plant was dedicated to making, testing, and characterizing basic active materials and then building them into small batches of cells, which are used for testing and future product development. Unlike Enerdel, in this plant A123 makes cylindrical cells, the format traditionally used in consumer electronics. Instead of cutting electrodes into book-shaped rectangles and then stacking the two electrodes and a separator into battery sandwiches, this process involves feeding the electrode-separator-electrode combo into a giant winding machine, which rolls that long sheet of battery material into individual "jelly rolls," each designed to fit into an aluminum or steel can. The result is a cylinder-shaped battery that looks like a slightly larger and more industrial-strength version of the common AA.

In a laptop battery, several of these little cylinders are wired in series to reach the desired voltage. Even the flat rectangular battery in your cell phone or point-and-shoot digital camera is most likely wound— just wound into an oblong shape and then placed in a rectangular container. Winding batteries is arguably faster and cheaper than the cut-and-stack process for making prismatic cells. The benefit of prismatic cells, however, is that all that easily accessible surface area makes them easier to cool, which is particularly important when you start making cells the size of a paperback book and loading hundreds of them into a car. The process for making prismatic cells is still expensive, though battery companies and carmakers say costs will drop as volumes increase.

Because A123 got its start making power-tool batteries, the majority of its output is still cylindrical, but this will change over time. In 2008 A123 bought the Korean company Enerland, which makes prismatic cells, and a new stimulus-funded A123 factory in Livonia, Michigan, will adapt the Korean process. The intent of the Livonia plant is to build prismatic batteries for customers worldwide.

As we left the small winding room, where sheets of electrode material are wrapped into cans for the purpose of building test batteries, we walked by a chrome-colored metal basin splattered with glimmering jet-black nanophosphate electrode slurry, like a bakery-grade mixing bowl splattered with high-test chocolate. Some janitor hadn't been doing his job. We paused to look at the mess, and Chiang shrugged. "Open one up, and everything in a battery looks like black ink," he said. He took a few more steps, then paused to clarify: "But it's black gold."

Back in Chiang's hacked Prius, we drove farther west along the Charles, bound for the small factory in the exurb of Hopkinton, where A123 assembles giant batteries that plug into the electrical grid. As we entered the freeway, I asked Chiang how a small company based in Boston could ever compete with the giant companies of Japan, Korea, and China. "That's a good question," he said. The short answer, he said, is that you have to aim for the moon. "You don't have much of an advantage unless you do something high risk," Chiang said. "To compete in incremental advancements is not a recipe for success."

It also doesn't hurt to manufacture in China. At the time of my visit,

A123 had five factories in China, all of which, coincidentally, were forty-five minutes from the Jiangsu Province village where Chiang's father grew up. Their presence in Asia dates from their initial deal to build batteries for Black & Decker, when it became "eminently obvious" that they would have to go to China.

It's not all about labor costs, Chiang said. Battery manufacturing isn't incredibly labor-intensive, anyway. Manufacturing in China is easier as much because of the speed with which you can build a new factory and get it running as for the lower labor costs. Chiang says A123's first electrode powder facility in China went from greenfield to production in only nine months. If you need machinery, steel, aluminum, you can get high-precision-grade specimens almost immediately. Chinese manufacturing has its drawbacks, however, notably the potential for intellectual property theft. "We ended up having to teach these guys how to make our state-of-the-art, world-class batteries," Chiang's cofounder Bart Riley told the *Chicago Tribune* a few months after my meeting with Chiang. "And some of them are [now] competing with us directly."

The new line of batteries that Chiang seemed most excited about were not, however, assembled in China. The individual battery cells were made in China, yes. But it's in Hopkinton where A123 builds the tractor-trailer-size grid batteries that could be an important lifeline as the company attempts to scale into a giant.

We arrived in an office park surrounded by pine trees and drove around back, where Chiang plugged his Prius into a charging station. Three shipping containers stood attached to loading docks, as if delivering shipments of Blu-ray players to Best Buy. These were A123's gigantic grid-scale batteries, each one in a different stage of assembly.

On the factory floor, Bud Collins, vice president for engineering in A123's Energy Systems Group, walked me through the various projects that can keep the company standing until building batteries for electrically driven passenger cars becomes a large and profitable business. He showed me a lithium-ion starter battery for a new supercar by a "well-known automaker." Next, the first and only FAA-approved lithium-ion aircraft battery, which Cessna uses to start the jet engines on Citations. Soon we arrived at a stack of giant white slabs that sit atop city buses in New York, Toronto, and San Francisco. Each was a 770-pound lithium-ion battery that turns a bus into an enormous Prius. They can produce 200 kilowatts of power on demand, which is enough, Collins said, that bus drivers have had to recalibrate their pedal-heavy feet. So far, they had shipped two thousand bus packs, which together had logged a total of four million miles.

Soon we were standing inside what Chiang said was the largest lithium-ion battery on the planet. A123 calls it the SGSS (Smart Grid Stabilization System). In this heavily modified fifty-three-foot-long shipping container, harsh fluorescent lights glowed overhead, and the white-noise roar of high-voltage current piled atop the hum of the cooling fans. Against one wall stood a supercomputerlike mass of battery—eighteen computer racks with eight trays per rack, and six battery "modules" per tray, and 96 individual lithium-ion battery cells per module, for a total of 82,000 cells per container. (The Chevy Volt's four-hundred-pound battery pack contains 220 cells.) It looked like some kind of portable cyberwarfare command center.

This was all impressive, but A123 faces challenges. The lawsuits with the University of Texas were still unresolved. The day I visited, Chiang was excited about a new deal with the plug-in-hybrid start-up Fisker Automotive, but Fisker had more doubters than any of the other major EV entrants. Enerdel had actually passed on the Fisker deal, because it would have involved investing in the company, and they didn't see the value in such an arrangement. (Enerdel already has an ownership stake in a car company—Norway's Think, a manufacturer of spartan, Euro-style electric microcars that are scheduled to arrive in the United States in 2011.) A123 took the chance, putting up to $23 million into Fisker in a combination of cash and stock. To some observers, this arrangement—and, for that matter, the Enerdel-Think deal—looked like pay-to-play. As one industry insider told me, this kind of deal made the American battery companies look desperate.

Well before the American battery stimulus, the Japanese, Korean, and Chinese manufacturers that rule the lithium-ion market were expanding their empires.

In May 2008, Sanyo, which at the time was the largest lithium-ion manufacturer in the world, announced a partnership with Volkswagen AG to build batteries for Volkswagens and Audis; Sanyo said it would spend $769 million by 2015 on the venture. That July, Panasonic announced that it was considering investing almost $1 billion to build a new factory in Osaka Prefecture that could produce fifty million lithium-ion cells a month. A month later, Mitsubishi, through a joint venture with the Japanese battery company GS Yuasa, would build a new lithium-ion factory in Shiga Prefecture. Nissan's joint venture with NEC, Automotive Energy Supply Corporation, declared plans to start building batteries for Nissan's upcoming electric car (the Leaf hadn't yet been officially announced).

During a 2008 talk at an electric-car conference in Washington, D.C., Charles Gassenheimer cited Nissan's recent purchase of twelve electrode-coating machines at a cost of some $150 million as evidence that this boom was serious. He said it was an encouraging sign that automakers were enthusiastic about electrification, and that this time, the electric car was not going to go away.

But Gassenheimer could also have seen Nissan's investment as an existential threat. In addition to the preparations in Japan, Chinese and Korean companies were arming themselves. In China, battery manufacturers in Shenzhen and Tianjin were increasing production, and some of them stated their intent to build cars too. Within a year, the Chinese government would make the conquest of the world electric car market an official state goal. In South Korea, not only was LG Chem planning to devote a factory to producing battery cells for as many as sixty thousand Chevy Volts a year, but Samsung was also launching an automotive joint venture with the German company Bosch at a cost of up to $400 million over the following five years.

The new factories that these companies were building were enormous, on the order of five hundred thousand square feet each, which is five times the size of the Enerdel cell plant I visited. China's BYD provides an excellent example of the scale that many Asian manufacturers have already achieved. The fifth largest lithium-ion manufacturer in the world, BYD's thirty thousand workers live in high-rise dorms on the company's four-square-mile Shenzhen campus. They work twelve-hour days. In early 2010, one industry analyst told me that BYD would soon inevitably be the world's fourth-largest lithium-ion supplier.

The colossus of the battery world was created in the fall of 2008, when Panasonic announced plans to buy Sanyo. Many analysts agreed that forming a battery juggernaut was exactly the strategy behind the merger. At the time, one Japanese analyst told *The New York Times*, "This appears to be the kind of deal where you add one and one and get three, instead of two. Their battery operations would truly be world-class." Including all their battery operations (not just lithium-ion), Sanyo sold more than $5 billion worth that year and had hybrid-battery partnerships lined up with Honda, Ford, and Peugeot Citroën.

"Battery manufacturers have realized that their marketplace gets bigger by like a factor of a thousand if we have electric cars instead of just handheld electronics," said Martin Eberhard, who had gone on to become a director of electric vehicles for Volkswagen-Audi. When I spoke to Eberhard in April 2010, he had just returned from a battery-hunting trip in Asia. "They realize that. So they realize that's a place to spend money. And I was in both Japan and Korea, and what I saw was great promise with many, many companies." He emphasized how un-settled the field is at the moment—few people even agree about the best shape for lithium-ion batteries. Small cylinders? Big cylinders? Prismatic cells? Vacuum-sealed pouches? "It's really the wild wild west," he said. "People are out there grabbing land while it's there to grab. What turns out to be the good land and the bad land is not clear yet."

"People always ask, 'Are we winning the battery race?'" James Green-berger said. "And I tell them: It's not a race, it's a boxing match. We didn't even show up in round one. Got knocked on our ass in round two. And we're about to be knocked down in round three and round four, because nobody's going to be able to outdo Panasonic or NEC or LG Chem. But the key to winning the boxing match is to stay in the fight long enough that you can pull out your left hook in the fifth and the sixth rounds. And in the United States our left hook is technology."

Yet the Asian companies are developing new technology too, com-mitting tremendous sums to research and development. And that means that extraordinary challenges lie ahead for American lithium-ion start-ups—particularly, Eberhard believes, A123. "I think A123's doomed,"

Eberhard said. "Their technology is no good. The energy density is too low, and you can't overcome it." He mentioned that Panasonic had recently announced that in 2013 it would begin selling new 18650 cells that hold 4 amp-hours at 3.4 volts. "That's like triple what you can do with an iron phosphate battery!" Eberhard argues that A123's original advantage, the inherent safety of its chemistry, is beside the point. For Eberhard, safety is a systems-engineering problem; what battery companies should be concentrating on, above all else, is increasing energy density and driving down cost. The companies with the most obvious advantage in the cost battle, of course, are the existing market giants, who have paid off their equipment and are already making billions of dollars a year.

The argument for dropping battery plants in the same neighborhood as the car companies they're supplying is straightforward: shipping these several-hundred-pounders across the world is a waste of time and money, and the distance introduces the kind of contingencies (dockworker strikes, trouble with Customs) that terrify procurement managers. But the argument that building batteries in the United States is a national-security issue is tougher to defend, and it involves weighing the relative advantages and disadvantages of generating American jobs and getting off oil as quickly as possible, even if the batteries that make that possible are imported. After all, imported batteries are not like imported oil. An advanced auto battery is a piece of high technology designed to last for years. Oil is a commodity we buy millions of barrels of each day and then burn for fuel.

Some national-security-focused Americans are fine with this. "If you care about, for example, climate change, your number one priority should be, how do I get those technologies into the marketplace quickly?" said Gal Luft, executive director of the Institute for the Analysis of Global Security, a Washington, D.C., think tank that focuses on energy-security issues. "Today—not in ten years, not in twenty. So if China does it, great. If Japan does it, great. If the U.S., even better. But I want to see it happen."

The key to making it happen is undoubtedly reducing cost. "Dollars per kilowatt-hour stored is all that matters," Eberhard said. "Let's picture an approximate Tesla battery. Let's say that it's 50 kilowatt-hours, it's

$20,000, and it weighs one thousand pounds, just for nice round numbers. Now I'll give you two choices. In choice number one you can have the exact same battery with five times the energy density, so instead of being one thousand pounds it now weighs two hundred pounds. The second choice is that you have the same battery but it costs one-fifth. Instead of costing $20,000 it costs $4,000. Which world would you rather have? In the first world, the Tesla Roadster gets to be a rocking car. It's a *really* nice sports car. In the second world, it's game over for gasoline."

For an industry cheerleader, James Greenberger is sober to the point of being a bit of a buzzkill. "In our mind the single greatest barrier to adoption of electric-drive and grid-balancing technologies is that electrochemical energy storage simply costs more today than do competing technologies that perform the same function," he said. "And until you solve that problem, a lot of the hoopla may not yet occur."

Greenberger likes to cite Geoffrey A. Moore, author of *Crossing the Chasm*, on the challenges of mass-marketing new technologies. "If you take a look at high-tech marketing and the experience of the high-tech industry, you know that the early adopter market for these transformative technologies is relatively easy to come by. We will sell EVs and PHEVs [plug-in hybrid electric vehicles] to folks who bought the Prius, there's no question about that. But that is not an economically sustainable market, and it's not a politically sustainable market, because if we find ourselves in five years with a PHEV and EV market that is entirely dependent on wealthy consumers and government subsidies, the government subsidies will go away. And so we have a fairly short time period— in my view, probably within five years—to figure out how we sell EVs to the general U.S. consumer, who is completely nonideological and not particularly interested in new technology. They just want a product that does something they do already and does it a little bit better."

For this to happen, automotive-grade lithium-ion batteries will have to get much cheaper. The most ambitious cost benchmarks come from the United States Advanced Battery Consortium: $300 per kilowatt-hour in a plug-in hybrid that can run for ten miles on electricity alone, and $200 per kilowatt-hour for a forty-mile plug-in like the Volt. Ted Miller, chairman of USABC and senior manager of energy storage strategy at Ford, explained the goals: "We want to produce plug-in electric vehicles as competitively as any other vehicle. That's the objective of $200." Two hundred dollars per kilowatt-hour is aggressive—an ideal to strive for—and Miller said he hadn't yet seen the technology that would make it possible, not even in the lab.

Still, that some of the biggest corporations on earth have invested billions of dollars in lithium-ion-powered automotive electrification suggests that, according to the kinds of internal calculations that carmakers and battery companies don't share, the math can work. "Last year we went out and did a benchmarking survey of lithium ion for various systems," said Dan Rastler, an energy storage analyst with the Electric Power Research Institute. "Vendor replies were all over the place, so we're going back to them and asking them to give more numbers. It's kind of a moving target." According to their bottom-up analysis, however, Rastler believes that there's no reason the large-format prismatic cells now going into cars won't eventually cost the same as the lithium-ion cells for consumer electronics that are today sold as a commodity. Rastler said that as of spring 2010, they were finding that lithium ion had already gotten down to a cost of about $600 per kilowatt-hour, essentially the same as a lead-acid module designed to do the same task. The results of a study by Paul Nelson and Danilo Santini at Argonne National Laboratory align with these estimates.

"Costs are coming down very fast," Yet-Ming Chiang said. "If you look at the pie chart of battery cost, there's no single thing that dominates." He said that hydrogen fuel cells are expensive largely because they use a platinum catalyst; platinum is pricey, and in hydrogen fuel cells nothing else can easily replace it. In lithium-ion batteries, on the other hand, "there are lots of things to cut."

Giant grid batteries like the ones Chiang showed me could actually help solve the fundamental scaling catch-22, which is this: Until they're built in massive numbers, electric cars will be too expensive for the majority of car buyers, primarily because of the cost of their exotic hand-built batteries. But they won't be built in bulk until there are hundreds of thousands of electrified cars on the road.

The economics are different for grid batteries than they are for cars. The electrical grid is so idiotically inefficient today that spending a small fortune on giant lithium-ion batteries to hook into the system could actually be a moneymaker. Chiang and Bud Collins say the only reason A123 got into the grid-battery business is that the global energy company AES called them up and asked for some. "This is all financially driven," Collins says. Enerdel is in the grid-battery market as well, starting with a deal to supply Portland General Electric with five 1-megawatt batteries that the utility will use for wind and solar power—to store electricity generated when the wind is blowing and the sun is shining so that it can be used anytime.

Silicon Valley veterans seem to be the most optimistic forecasters of battery cost, which isn't surprising considering the shrink-to-nothing economics of computing. The absolute lowest possible cost of a lithium-ion battery is called the cost floor, and that is something that Martin Eberhard says does not exist. "I don't know what a cost floor is," he said. "That's a concept I don't believe in. I have seen cost floors—absolutely cannot get cheaper than X—presented by many a company, and then I can go to company Y and show you a price that's already below that."

Once the research and development and the factories and machinery are paid for; after the cost of labor and shipping, of keeping the lights on and the water running—eventually cost comes down to raw materials. And as the battery boom began, concerns about the availability of lithium, which had never before been mined or traded in significant quantities, raised an entirely new set of questions about the inevitability of an electric-car age.

9

THE PROSPECTORS

In December 2006, an energy analyst named William Tahil posted a paper online titled "The Trouble with Lithium." In it, he argued that basing an electric-car revival on the lithium-ion battery was nothing but a headlong rush into dependence on yet another finite resource, an addiction to oil traded for an addiction to lithium. According to Tahil's analysis, lithium reserves were dangerously limited; there was nowhere near enough economically recoverable lithium in the world to support a global switch to lithium-ion-powered electric cars. "If the world was to exchange oil for Li-ion based battery propulsion," Tahil wrote, "South America would become the new Middle East. Bolivia would become far more of a focus of world attention than Saudi Arabia ever was. The USA would again become dependent on external sources of supply of a critical strategic mineral while China"—home to significant lithium deposits—"would have a large degree of self sufficiency."

Earlier that year, Tahil had published another paper. That one was called "Ground Zero: The Nuclear Demolition of the World Trade Centre." He argued that two "clandestine" nuclear reactors, buried some 250 feet below the World Trade Center, were deliberately melted down, Chernobyl-style, at the same moment the hijacked airliners hit the Twin Towers on September 11, 2001. No "let it happen on purpose" for Tahil. Not even the kind of controlled implosion that commercial firms use to take down buildings could explain to Tahil the events of that day. Instead: "The evidence is overwhelming and incontrovertible that the Twin Towers were subjected to far more than just a conventional Controlled Demolition. They were each pulverized to dust by a Nuclear Explosion." Who buried these secret nuclear reactors beneath the World Trade Center? Who sabotaged them that morning? Tahil did not claim to know. But, as he pointed out, the first atomic bomb test was called Operation Trinity, and "It is interesting to note that the church at the WTC was called Trinity Church."

This didn't damage Tahil's credibility. At least not right away. Numerous media outlets seized on Tahil's warning about "peak lithium," and soon the public was being bombarded with the notion that because of a critical resource shortage, the electric-car revival was doomed before it even began. Tahil was quoted in mainstream publications like *Forbes* and industry magazines like *Chemical Week*. Every newspaper or magazine story on the lithium-based electric-car revival suddenly seemed to contain a boilerplate warning from Tahil on the consequences of dependence on lithium-ion batteries.

Each of those stories also tended to contain a rebuttal from the geologist R. Keith Evans. "It was total bullshit," Evans told me. We were walking down a hallway at Caesars Palace in Las Vegas, where I had intercepted him as he headed to the casino for a cigarette. I had come to the second annual Lithium Supply and Markets conference looking for the truth behind the lithium-supply debate, because even though Tahil had turned out to be a little unhinged, that didn't necessarily mean his math was wrong. Plenty of credible people had also been expressing unease about the reliability of world lithium supplies. Donald Sadoway, for example, a proud ambassador for electrochemistry, had soured on lithium because of such concerns. "I started asking, how much lithium is there on the planet that's readily available?" he said. "Where are the decent lithium resources? They're in surface brines on the border between Bolivia and Chile and in China. It's not the Middle East, but it's not Utah. So I started thinking—are we going to trade in importation of petroleum for importation of lithium? And what's to stop the people that have lithium from arbitrarily raising its price, so we go from OPEC to LiPEC? Or the Organization of Lithium Exporting Countries—OLEC?"

There, on the long escalator down to the casino, Evans explained how he became involved in the peak lithium debate. I hadn't asked how the feud started. I had actually asked him why Toyota, a rich source of anti-lithium-ion talking points in the months after the Chevy Volt unveiling, had recently taken an ownership stake in a minor lithium producer called Orocobre. But for Evans, the story began with Tahil. Once Tahil's paper grabbed the public consciousness, Evans suggested, every major business contemplating a future involving large-scale purchases of lithium probably told their procurement offices and analysts to look into the situation and, if necessary, to guarantee a supply.

It all happened quickly. Two years after the Volt's debut, the London-based trade magazine *Industrial Minerals* convened the first Lithium Supply and Markets conference in the lithium capital of the world, Chile. A year later, by the second meeting, Toyota had come around so strongly that they were now part owners in a company whose mine wasn't even operating yet, whose only selling point was a claim to a promising salt flat in the far north of Argentina.

The doomsaying about lithium supplies continued, despite Evans's best efforts. When Tahil published "The Trouble with Lithium," Evans, who had spent nearly forty years studying lithium supplies—first, in the early 1970s in Zimbabwe, and then during the false start of the late 1970s in the Salar de Atacama, the world's largest and purest active source of the mineral—was apoplectic. "And so I decided that I had to come out of retirement."

In response to Tahil, Evans wrote his own estimate of the world's lithium supply, a white paper called "An Abundance of Lithium." In it, he told of an urgent conference that the U.S. Geological Survey held in 1975 to warn of an impending shortage of lithium from the year 2000 on—for use in nuclear fusion reactors. That scare generated the first serious estimate of global lithium supply. In the years after the 1975 conference, data on lithium abundance multiplied, and when Evans wrote his reply to Tahil, he explained that the current global lithium deposits amounted to some 28.4 million tons of lithium (or 150 million metric tons of lithium carbonate, the most common form in which lithium is produced and sold). The annual demand for lithium was 16,000 metric tons. Particularly after taking into account the rush of exploration and research that was accompanying this burgeoning lithium boom, there seemed to be little to worry about.

The fencing continued. Tahil hit back with "The Trouble with Lithium 2: Under the Microscope," this time addressing Evans by name and accusing him of confusing the abundant theoretical supply of lithium contained in Earth's crust with lithium that was economically extractable.

Evans lost his patience. He posted a response paper online that was part lithium-reserve update—estimating upward based on new exploration and information—and part direct attack on Tahil.

Estimated global lithium reserves and resources are increased slightly from the earlier figure to 29.9 million tonnes Li.

This revision is written in response to a recent report which is alarmist in its gross underestimate of resources and, in several respects, ludicrous.

That was the entire abstract.

Nonetheless, governments, brokerage houses, automakers, and electronics manufacturers were all in Vegas that week, all of them eager to understand the supply of the mineral that suddenly appeared as if it would be essential for energy storage for decades.

At the cocktail reception the night before the conference began, five showgirls in full peacock-feather undress danced to "Viva Las Vegas" before a Tesla Roadster, two electric motorcycles, and a DeLorean that had been converted into an electric car. The crowd was young, predominantly male, and had come prospecting from Japan, South Korea, France, Germany, Switzerland, Italy, Chile, Bolivia, Peru, Argentina, Brazil, Mexico, Canada, Australia, Norway, Serbia, the United Kingdom, and various American states.

Around a standing table in the middle of the convention center meeting room, I talked to two employees of the U.S. Defense National Stockpile Center, one of them a thirtyish guy who confessed to being gleefully ignorant about the lithium markets and lithium-ion batteries in general. He had no idea that the lithium-ion batteries that would soon go into electric cars were made of a different combination of elements from the batteries in his cell phone. I asked whether he would be attending the "field trip" at the end of the week, a charter flight to northern Nevada to visit the site of what could soon be the only large-scale lithium mine in the United States—a source of lithium in the lower forty-eight large enough that, according to the company developing the property, it could theoretically supply the entire world. He was not. "Are you kidding? The government is paying for this." (The field trip cost $1,350.)

The next morning I ran into Defense National Stockpile in the maroon-carpeted hall just as the opening presentation was beginning. He looked rough. Vegas. "I was out *laaate*," he told me conspiratorially. "That's why I've got the coffee *and* the Gatorade."

"The biggest weakness we have in our system is policy makers' ignorance," said the opening speaker, Gal Luft of the Institute for the Analysis of Global Security. "Go to the records: When was the last time the word 'lithium' was mentioned in Congress? Never. Government officials know absolutely nothing about this. They know about as much as we know about the ingredients in our food. They don't know and they don't care."

Luft was not arguing that a lithium shortage was likely. He was actually arguing that a shortage of everything else that goes into an electric car—particularly rare-earth metals, which are essential for building the permanent magnets that drive electric motors but which are mined almost exclusively in China—was more likely. Lithium was the good news. Still, lithium is, as he put it, "the yeast in the dough." If we're going to put one million electric cars on the road by 2015, we need to make sure we have the raw materials to do so, and so far we haven't done that, he argued. As he put it to me later, the United States is like "a bakery that doesn't take inventory."

The lithium industry has historically been tiny and comprehensively ignored by mining companies, which are generally interested in more captivating resources like gold, silver, and uranium. But since the launch of the Volt in 2007, the number of companies and countries scratching their way into the lithium business has become overwhelming. "There are sixty-seven companies listed on the TSX [Toronto Stock Exchange] alone, and over one hundred companies globally looking for lithium," the geologist and market analyst David L. Trueman said in a presentation the first day of the conference. In Canada, the exploration situation had gone "berserk," Evans said in his presentation, beaming an illegibly dense list of Canadian exploration projects onto the presentation screen. Beyond that, a company called Nordic Mining in Finland was gearing up to extract lithium from a deposit of a rock called pegmatite. In Australia, the companies Galaxy Resources and Talison were preparing to do the same. A secretive start-up called Simbol, staffed by former Lawrence Livermore scientists, claimed to be able to meet some 20 percent of the world's lithium demand by filtering the water that flows through four geothermal plants on California's Salton Sea. Western Lithium was sitting atop an enormous deposit of lithium-rich clay in northern Nevada. The mining giant Rio Tinto was exploring a deposit of a lithium-bearing mineral called jadarite in Serbia. In addition to the recently announced Toyota-Orocobre deal, Toyota had just formed a joint venture with the South Korean steel company POSCO, and the Canadian auto-parts supplier Magna had also made investments in lithium exploration companies, according to an analyst with the London research firm Roskill.

In Las Vegas, the Oligopoly, as I heard the three companies that supply the vast majority of the world's lithium referred to more than once, had an important message for the prospectors: Our company is sitting on enough lithium to supply the world for hundreds of years to come. You junior mining companies and start-up dreamers in the audience, do not think that you can compete.

The first to deliver this warning was Sociedad Química y Minera de Chile, or SQM. Patricio de Solminihac, the executive vice president and COO of SQM Chile, tried to discourage any potential wildcatters by explaining that its source of lithium, the Salar de Atacama, a 280,000-hectare salt-encrusted depression high in the Atacama desert, had insurmountable advantages: It contained forty million tons of lithium carbonate equivalent, the largest known commercially exploitable reserves in the world. Its brines contained a higher concentration of lithium than any other such deposit. That brine was low in the ancillary minerals that eat away at recoverable lithium content. The Salar de Atacama was close to the port city of Antofagasta, a Pacific coast mining hub. In 2009, SQM produced forty thousand tons of lithium carbonate. Count up the current expansion plans of the existing lithium-market players and the world would be well supplied for the next fifteen years. If any new entrants got in the business, the market would be oversupplied for the next twenty years.

Next, the German company Chemetall gave a similar presentation, ending with a coup de grâce that drew some mockery at the afternoon coffee break: a mention of its ability to reopen a dormant mine in Kings Mountain, North Carolina, which is full of the lithium-containing rock spodumene. "That's like rolling out your ninety-five-year-old grandmother in her wheelchair for extra numbers," I heard someone say.

The third member of the oligarchy, FMC—a century-plus-old chemical company originally called Food Machinery Corporation, which in 1985 got into the lithium business by buying the Lithium Corporation of America—rounded out the discouragement by stressing its credentials not as a mere mining company but rather as a manufacturer of high-quality custom chemicals. "The automotive market has some of the toughest quality requirements in the world," said Jon Evans, the manager of FMC's lithium division, and FMC's products—an array of processed lithium powders and ingots and solutions with names like LectroMax and LectroLyte—already met those specifications.

During a panel that afternoon, an analyst offered a helpful tip for interpreting the day's presentations. "We as listeners—we're hearing competing agendas," said Eric Zaunscherb, a mining analyst with the Vancouver-based finance firm Cannaccord Adams. "The big three, the Oligopoly, are sitting on one shoulder, and juniors are sitting on the other. We have competing agendas, and it's important to remember that as we listen to people tell us how many tons of lithium are out there."

Even so—even assuming that the three members of the Oligopoly were using their most wishful measurements and exaggerating the amount of lithium they had access to—the competitiveness of this market, the number of new projects, the amount of new exploration made concerns about an impending lithium shortage seem seriously unfounded. Particularly because lithium isn't burned for fuel like oil is. It's used to make a machine that stores energy, much like copper or iron or any other metal.

Each member of the Oligopoly runs at least one major operation in the so-called Lithium Triangle, the corner of South America that is to this element what the Middle East is to oil. Ten thousand years ago, a series of saline lakes stood on the Andean plateau where the nations of Bolivia, Chile, and Argentina now intersect. When those lakes evaporated, they left behind enormous *salares*—deep, solid lakes of salt crystal sitting in basins surrounded by volcanoes. The volcanoes had long been factories for rocks high in light elements like magnesium, potassium, boron, and lithium. Over the millennia, as the snow capping those volcanic peaks melted each year, water glided downhill through that rock, leaching minerals free and dragging them down into the basin, where the resulting brine soaked into the salt. As a result, this remote region is today the world's richest source of lithium.

SQM extracts approximately 30 percent of the world's lithium supply from a single salt flat in the Lithium Triangle, Chile's Salar de Atacama. A Chemetall plant sits on that same salar. FMC drains most of its lithium from a similar salt flat across the border in Argentina, the gothically named Salar del Hombre Muerto.

To the north of FMC's source is the country that may be richest in lithium of them all: Bolivia. Its Salar de Uyuni is thought to contain the world's single largest lithium resource, estimated at some 8.9 million tons. Because of poverty and political instability, that resource is untapped. But early in the electric-car revival, the Bolivian government committed to developing it, and by the time of the conference in Las Vegas, the Salar de Uyuni had gone from an obscure natural wonder to the investment opportunity of the century. Which suggests another way of interpreting the puffing and strutting from the existing three major producers: They aren't fighting off competition only from the little guys. They're fighting it off from the biggest resource in the world, located just across the border from their most fecund mines.

The evening after the close of the conference, Western Lithium invited me to a "casual dinner" at Guy Savoy, the double-Michelin-starred American outpost of the Parisian chef's empire. Vegas has gone upscale in recent years, but this restaurant still seemed entirely out of place in the smoky, tawdry halls of Caesars Palace, with soaring ceilings, argon-filled wine cabinets, and obsequious French waitstaff who seemed to magically appear when any glass approached empty.

After two days in the bland, equalizing environment of a conference hall—everyone wearing the same plain business wear, everyone drinking the same squat half bottles of Diet Coke—walking into the predinner gathering around the bar at Guy Savoy felt like stumbling into a James Bond movie. Meet Yuhko Grossmann, an elegantly dressed woman of ambiguous nationality who was charming the group of six or seven strangers like an experienced, wealthy hostess. Upon meeting her husband, I realized that that is what she is. Meet Ed Flood. Ed Flood is a trim, handsome man in his early sixties with long, coiffed silver hair and a touch of a Reno drawl to his speech. His suit screamed money. His aura screamed money. I've been in elevators with guys like this; they usually own the company I work for.

We were ushered into a private banquet room, and Flood took the head of the table. Grossmann sat in the center. The amuse-bouche arrived: two tiny fois gras and black truffle sandwiches on skewers.

Ed Flood, who was not at the conference, and whose name I had heard for the first time the day before in a presentation by Western Lithium president Jay Chmelauskas, was our host. Flood is the chairman of Western Lithium, a position he took when Western Lithium was spun out of Western Uranium, where he was the founding president. Additionally, Flood is managing director of investment banking for Haywood Securities, a London-based arm of the Vancouver finance firm. But Flood made his real money at Ivanhoe Mines Ltd., the international mining firm where in 1994 he became founding president and where he still sits on the board of directors. Based in Vancouver and traded on the New York Stock Exchange, Ivanhoe is now best known for developing, in coordination with Rio Tinto, the Oyu Tolgoi project in southern Mongolia, one of the largest copper and gold mines on the planet.

The head waiter came into the room and demanded attention. The next course would be one of those that made Chef Savoy famous in Paris many years ago: a mosaic of milk-fed poularde, foie gras, celery root, and black truffle jus. It came out looking like the most expensive slice of pimiento loaf in the world.

Grossmann lived in Monaco, near the casino, she told the man sitting next to her. She moved there nine years ago; it's much more of a small town, more of a true community, she said, than its reputation as a haven for international oligarchs might suggest.

The head waiter announced another course that made Guy Savoy famous: water-based artichoke and black truffle soup, topped with floating shavings of black truffle and parmesan and paired with a mushroom brioche roll served with black truffle butter. When the waiter finished his exposition on the dish and left the room, all conversation was silenced.

"Yuhko and I went truffle hunting in Piedmont," Ed Flood told the table.

Sitting across from me was Oscar Ballivian, a Bolivian geologist and consultant for the French industrial giant Bolloré, which had been sniffing around Bolivia's lithium reserves for some time. In 1981, Ballivian and a partner became the first scientists to measure the lithium levels of the Salar de Uyuni. Since then he has consulted for what seems like every corporation or government that has ever had any reason to be interested in lithium. I told him that I'd like to come visit Uyuni, and he gave me his card and offered to help. "We don't get meals like this in Bolivia," he said with a wink. "If you e-mail me and mention this dinner, I will remember."

Ballivian may not get meals like this in Bolivia, but he surely gets them elsewhere. He came to Las Vegas from Barcelona, and before that he had been in Paris. Last year he traveled to Paris five times in his role with Bolloré. He spoke self-deprecatingly on behalf of his country, sighing while explaining the difficulties of exploiting Uyuni's lithium motherlode. Bolivia's president, Evo Morales, insists on developing the resource without outside investment, and according to Ballivian, the government has no idea what it is doing. Unless it brings in outside technical expertise, it's not going to happen. Bolivia is one of the poorest countries in the western hemisphere, and the Salar de Uyuni is located in one of the most remote and unforgiving corners of that country, at an altitude of twelve thousand feet above sea level. Bolivia is landlocked; diplomatically, the government is tighter with Chávez in Venezuela than with Obama in America.

"The only problem with Bolivia is Bolivianos," Ballivian told me. I wasn't sure whether he was referring to the currency or the citizens, but he followed with a clarifying joke. "God is making Bolivia," he began, "and he says, 'We're going to give this country all the copper, all the tin, the fruits, the vegetables, the animals . . .' Someone says, 'Are you sure you want to give all this to one country?' And he says, 'Well, wait until you see the human beings we're going to put there!'"

After beef tenderloin topped with a tripe-white blob of bone marrow, and dessert, we were ushered out. We said good-bye to Ed Flood and Yuhko Grossmann, who would not be visiting the lithium mine site with us the next day. Yuhko had already seen it from a helicopter, and Ed was visiting his mother in Reno. The rest of us would rise before dawn to catch a charter flight north.

I walked downstairs and out to the front of the casino with a fashionably suited Canadian mining analyst, whom I asked, What's the deal with Ed Flood? "He's a guy who had a few big plays, a few big wins." He *played*, and he *won*. Regarding the extravagance of the dinner, the talk of Monaco, the tales of truffle hunting, he informed me simply, "That's what you get with the mining industry."

The next morning, nineteen mining industry types and I gathered in the casino to catch a bus to the airport. It was cold, and as I stood chatting with an employee of the large French metallurgical firm Eramet, a sunrise the color of glowing sandstone shone down on the fake Eiffel Tower in the distance. A Santiago-based marketing rep from SQM would be visiting Western Lithium that day, as would assorted other investors, analysts, the man in charge of the U.S. Geological Survey's annual report on lithium, and various international seekers of a steady supply of the mineral: scouts from the major Asian electronics corporations, a Japanese brokerage house, the Korean steel firm POSCO.

At McCarren airport, the casinos and hotels of the strip looming in the distance, we walked out onto the tarmac and boarded a chartered prop plane. "Up to beautiful Winnemucca," our pilot said drily. (He would be spending the day at beautiful Winnemucca's minuscule airport.) We flew northwest along the Nevada-California border, skirting the bombing ranges and UFO test grounds of central Nevada. Below, the corrugated rust-and-brown landscape was dusted with snow. To our left, Death Valley. To our right, Yucca Mountain and Area 51.

Halfway through the flight, our tour guide, a Western Lithium geologist named Dennis Bryan, announced that we were beginning to approach Chemetall's brine-based lithium operation in Silver Peak, Nevada. The crowd rushed to the windows like boys on a schoolbus gawking at a passing Ferrari. (Chemetall does not seem to welcome the new competition in the neighborhood. The day before, I had asked Monika Engel-Bader, the husky-voiced German president of Chemetall, what she thought of Western Lithium's prospects. "They don't have a chance," she said wearily while lighting a cigarette.)

From the Winnemucca airport we boarded a bus for an hour's drive north. We left that small mining town of bad motels and sad casinos and crossed the Humboldt River, the body of water that the forty-niners followed west in search of gold. There is still gold in these hills—some of the largest gold-mining companies in the world operate out of Winnemucca—but the days of striking arm-sized veins are gone. Now, gold mines are generally open-pit operations that chemically rake flecks of "no-see-um gold" from what appears to be, because it is, dirt.

Bryan was an enthusiastic tour guide. Over the bus intercom he explained the centrality of mining and gambling to his home state. Gaming and mining, respectively, rank as Nevada's top two industries. Mining is actually why this place was granted statehood; during the Civil War, Abraham Lincoln needed the silver from Nevada's Comstock Lode to finance the Union Army, and so in 1864, this land of gambling and mining and gambling big on mining became the thirty-sixth American state.

Twenty million years ago, Dennis told us, this region was downright Hawaiian in its volcanism. Volcanoes leave behind interesting stuff when they die: gallium, uranium, gold, mercury, potassium. Lithium, of course. We were headed to the McDermitt Caldera, an ancient collapsed volcano, a vastly older and more eroded version of the natural phenomenon most commonly associated with Oregon's Crater Lake. After the volcano's collapse, a geological phenomenon called basin-and-range faulting began, drawing north–south-oriented swaths of land down, toward the center of the earth, leaving mountain ranges on the edges and generally rendering the landscape unrecognizable (except to a geologist) as the site of once-violent volcanic activity. Ten thousand years ago, this landscape—north-trending mountain ranges intercut with sagebrush plains, some of which have been plowed and irrigated and implanted with alfalfa and potatoes—lay beneath a lake that stretched from Reno to Oregon.

In the 1970s, Chevron Resources was searching for uranium in the McDermitt Caldera when they detected an "anomalous clay lens" that was extraordinarily high in lithium. This, of course, was during the short-lived alternative-energy frenzy. Chevron drilled holes, tested core samples, explored, mapped the land in detail—and then the 1970s ended, recession hit, Reagan became president, oil became cheap again, and sitting on the rights to a giant lithium deposit no longer meant a damn thing. "They were thirty years too early," Bryan said. In 2005, Western Uranium took over the Chevron claims, which Chevron had sold in 1991 to another company, which subsequently let the claims expire. In 2007, Western Uranium spun those claims out into a separate company, Western Lithium. The land beneath which this lithium-rich clay resides is, like more than 80 percent of the state of Nevada, federal land, but Western Lithium currently holds two thousand mining claims on it, each one covering twenty acres.

In the geological order of things, lithium is what the Princeton geologist Kenneth Deffeyes described to me as a "misfit." When the lava from northern Nevada's volcanic days began to cool and harden into granite, a number of common elements made their way into freshly baked granite, but water and a medley of minerals including lithium were left out. The "leftover juices," as Deffeyes calls them, grew into veiny, crystal-rich rocks called pegmatites. The particular pegmatite that happens to be high in lithium is called spodumene (which is what Talison and Galaxy Resources are mining and purifying into lithium carbonate in Australia). Sometimes, these rocks get melted again; when they cool, they can form a rock called rhyolite. A particular strain of rhyolite that's rich in sodium, potassium, and lithium happens to be found near a number of major lithium deposits, including Bolivia's Salar de Uyuni and the Western Lithium site we're visiting today. The theory is that over time, the water table leaches lithium out of rhyolite and it moves around underground, eventually settling in a comfortable spot. That Western Lithium's deposit is bound up in clay makes perfect sense. "Lithium loves clay," Deffeyes said.

We drove by a crumbling, forgotten gas station that looked like a stage prop from a film about the postoil apocalypse. Tumbleweeds should have been rolling by. Chmelauskas, the company president, grabbed the microphone from Bryan and proclaimed that one day, all gas stations will look like that. "Our kids will say, 'What's that, Dad? What's a gas station?'"

We turned into a muddy driveway and parked next to the rented ranch house that the company uses for its field headquarters. The yard was lined with shipping containers filled with excavated earth from the mine site. In the backyard shed, core samples covered plywood tables. This is how mining companies determine what they're sitting on: pull up long cylinders of earth and map the composition of the underground by systematically analyzing each inch of the sample. By the time of our visit they had determined that between ten to fifteen feet of "overmatter" and a bedrock foundation of rhyolite rests a 300- to 360-foot-thick slab of lithium-rich clay known as hectorite. In spots this slab is intercut with stashes of ancient ash that is worthlessly low in lithium, but overall, the numbers are good, with the clay averaging an economically viable 4,000-plus parts per million (ppm) lithium.

Standing next to the sample-covered tables I talked to Chmelauskas, whose last project was the undeniably less virtuous task of overseeing the construction of the largest open-pit gold mine in China. "Now I wake up every day and I'm saving the world," he said.

It's his job to be this way, but he talked about the impending electric-car revolution like a man who had seen God. Or hell. By which I mean the air quality in China. "When I lived in Beijing, my family had what we called a building index: How many buildings can you see when you look outside?" he said. "Most days it was one." Two days earlier, during his conference presentation, he had made a similar case based on personal experience. "I was born in 1969, when the population of the world was three billion. Now it's seven billion. In my lifetime it's expected to go to nine billion." Those exploding populations don't just want to get by; they want to have the same quality of life that we have in the West, Chmelauskas argued. They want to go cruising through the grasslands of Mongolia on vacation. "We've got depleting resources, expanding populations, and expanding expectations from those existing populations," he said. Those factors, along with China's close brush with death by pollution, are driving the Chinese government to mandate electrification, and in this case, as goes China, so goes the rest of the world. "If China goes electric, GM better have electrics, or they're not going to sell cars there," Chmelauskas told me.

We put on rubber boots and weatherproof jackets, loaded into 4 × 4s, and left for the mine site. We drove west toward the Jackson Mountains, a low range separating the agricultural valley that contains the small town of Orovada from the caldera, the remains of that ancient, fertile volcano. This is antelope country, coyote country, desert bighorn sheep country. The road was smoothly paved, and when we took a right onto a dirt lane, driving only a few hundred yards up a sagebrush-covered hill, I was a little taken aback by the location of the slice of earth that Western Lithium says is capable of supplying the world's current lithium needs all by itself. It's a random hill in the empty American West, an unremarkable, sagebrush-covered mound in a relatively scenic bit of American mountain country.

We took a right on another dirt road. Ahead of us for maybe a quarter mile, to our right and our left, to the top and bottom of this ridge, and then behind us another several hundred yards—that's where the clay is. Underneath us.

We stopped at a fresh gash that trucks and shovels had gnawed into the earth and walked down into the trench. It was not violent mining. As promised, the faces on either side of the cut revealed fifteen or so feet of dirt, and then clay. My boots bounced slightly on the clay, as if we were walking on a giant sheet of Play-Doh. Chunks of hectorite lay scattered across the intact clay floor, wet to the touch and roughly the liverine color of the fois gras in last night's mosaic of milk-fed poularde.

Western Lithium says this giant clay sponge contains the equivalent of five hundred thousand tons of lithium carbonate, theoretically enough to satisfy the current world demand for four years. Two investors in the company, young brothers running a family fund in San Francisco, milled around beside me. I asked them what they thought. "I like what I'm seeing," one said. "Other open-pit mines we've been to, it looks like they had to move the whole fucking world to get in there." The other brother: "I see cash flow, is what I see."

In two to five years, the ground beneath our feet would be an open-pit mine. If that one isn't enough, there are four more clay deposits to our north that can be opened up. Earlier I had asked Chmelauskas what the environmental drawbacks to a mine like this would be. He was frank. Mining lithium from clay makes much less of an impact than many other extractive industries—there are no toxic chemicals involved, and no blasting (one person described the method of mining employed here as "gardening")—but mining is never impact-free. "I mean, we're going to put a big hole in the side of that mountain," Chmelauskas said. "But you have to weigh the net costs." Meaning, yes, we're going to gut one hill in a remote corner of Nevada, but isn't that a sacrifice worth making in order to secure a major North American source for the active ingredient in the petroleum-free cars of the future?

When the mine goes live, the clay from this hill will be crushed, calcined, then mixed into a liquor of sulfates. After that comes the water leach, then the addition of soda ash, and then from that solution lithium carbonate should precipitate out. When the mine is running at full scale, it will produce 27,000 tons of lithium carbonate a year, along with 115,000 tons of potassium sulfate as a by-product. It'll cost 89 cents a pound to process that lithium, and Western Lithium should make $263 million a year.

On the flight back to Las Vegas, I sat across the aisle from Oscar Ballivian. I leaned over to ask him whether Bolivia or Western Lithium—two projects starting from essentially the same stage of preliminary exploration—was more likely to start producing first. He answered by listing the many companies he had consulted for since beginning his research at the Salar de Uyuni in the 1980s. Negotiations with the government had never gone far enough to get Uyuni into production. Now, Ballivian said, Morales insists on developing Uyuni himself, in order to keep the Bolivian people from getting robbed, as so many South American people have by so many foreign mining concerns before. There's a new mining minister, and he might be accommodating, but who knows, Ballivian said. There are many problems. You can see for yourself, if you go down there.

10

THE LITHIUM TRIANGLE

The unaccustomed human brain has a hard time processing the Salar de Uyuni. The closest analogy for this endless expanse of white substance is the ice sheet covering a frozen lake, and so the first time you drive onto the salar, something deep in the spine warns you that at any moment the surface could split, and you could plunge in.

The salar, however, is solid, composed of an unfathomably large amount of solid material—forty-seven billion cubic meters of salt and gypsum and brine and mud and fossilized brine-shrimp feces. It is also a sponge for the brine, known in South America as *salmuera*, that is perhaps the world's richest source of lithium.

During the kind of overcast, rainy weather that welcomed me during my first moments on the salar, all recognizable nonsalt landmarks disappear in a fog of white and gray. In every direction, including up and down, there is only the dull white of salt and cloud. Under my feet on the day I arrived at the small and desperate operations of the fledgling Bolivian lithium initiative was a dam made of bulldozed salt, the main wall for a series of evaporation pools. If all goes as planned, one day this pool and others like it will be essential for processing the world's largest supply of this mineral.

As I would soon learn, however, it could be exceedingly difficult to make that happen.

Since Oscar Ballivian published the first rigorous measurements of the Salar de Uyuni's lithium content in 1981, geologists have known about the rich mineral stash buried beneath the salt. Until recently, however, no one cared all that much, because lithium was a niche product, nothing that could compete in the minds of speculators and investors with the likes of gold, silver, uranium, platinum. In the early 1990s, the American company Lithco attempted to get a mining concession from the Bolivian government, only to be beaten back by resistance from the locals, who even then, when lithium was something hardly anyone ever thought about, were afraid of foreign companies coming in to steal their mineral riches. But only after the announcement of the Chevy Volt and the emergence of the "peak lithium" scare did the frenzy begin. William Tahil's warning—that Bolivia would become a greater source of world attention than Saudi Arabia had ever been—made an irresistible story, particularly given Bolivia's long, traumatic history of natural resource exploitation and its fascinating and tumultuous recent politics. Suddenly, starting in 2009, the Salar de Uyuni was in the news—in *The New York Times*, in *Le Monde*, on ABC, the BBC.

Named after the great South American liberator Simón Bolívar, Bolivia is one of the poorest countries on the continent. Bolivia's gross domestic product in 2009 was $45.11 billion, roughly the same as ExxonMobil's profit the year before. Since gaining independence from the Spanish on August 6, 1825, the country has experienced some two hundred coups, countercoups, and other governmental spasms. Much of the country's internal conflict comes from a geographical and ethnic divide between the western Andean highlands—home to the majority of the country's 60 percent indigenous population—and the eastern lowlands, the Amazon basin territory where the citizens are predominantly white and mestizo. Among the indigenous people, 30 percent speak Quechua, the language of the Incas, while 25 percent—including the country's president, Evo Morales—speak Aymara, the language of the people who ruled before the Incas took over around A.D. 1450.

As Oscar Ballivian told me in the form of his joke about the divine folly of Bolivia's creation, this is a nation of vast natural resources. Those resources, however, have historically been plundered by foreign interests without contributing anything to Bolivians. When the Spanish conquered the Incas, one of their first acts was to drain the rich veins of Andean Bolivia of their silver. They did so using enslaved natives, hundreds of thousands of whom died in the mines. The most notorious example is found in the Bolivian city of Potosí. Potosí sits beneath a mountain that is called Cerro Rico for its once staggeringly productive silver reserves. In the early seventeenth century it was one of the richest and most important cities in the western hemisphere, a place whose name was shorthand for unimaginable riches. When Paris and London were squalid middle-size cities whose main export was bubonic plague, Potosí was like Dubai, a desert city made of money. Now Potosí is a poor town in a desolate and forgotten corner of the world, a rough place where coca-chewing prepubescent children work every day chipping at the old mining tunnels that perforate Cerro Rico, looking for any shards of silver the colonial era left behind.

Bolivia sits atop South America's second largest natural gas reserves, after Venezuela, but so far it has failed to develop those resources, and the story of its failure to do so says plenty about the relationship of ordinary Bolivians to extractive resources. The seeds of the Bolivian "gas wars" were planted in 2002, when the president, Jorge Quiroga, an American-educated self-described yuppie, began considering a plan to build a $6 billion pipeline to send Bolivian natural gas to Chile, where it would be processed, liquefied, and sold to Mexico and the United States. Because it involved two of indigenous Bolivians' least favorite countries—Chile and the United States—and because it would apparently deliver only a fraction of the true value of the gas to Bolivian interests, the plan was deeply unpopular.

The gas pipeline was still being considered during the 2002 election, which was riven by social movements that would transform the country. A charismatic coca-grower's union leader named Evo Morales, the presidential candidate for the Movement Toward Socialism (MAS) Party, was making a surprisingly strong showing against Gonzalo Sánchez de Lozada, or "Goni," the former president and a staunch free-marketeer.

The Bush administration was vocal about its distaste for Morales and the MAS, which was a mistake. Three days before the election, the American ambassador raised the possibility of cutting aid to Bolivia if Morales won, and the Morales vote immediately surged, putting him a close second to Lozada. (Morales took to calling the U.S. ambassador his "campaign manager.") Lozada and Morales were forced into a runoff election. Lozada won, but barely, and the result was a fragile coalition government.

Officially, Lozada remained undecided about the gas pipeline, but all the while he was pursuing the Chile plan. The public hadn't forgotten about the pipeline. It had come to represent every bit of wealth that had been stolen from the country over the centuries. On the surface, the gas pipeline debate was about the share of profit that Bolivia would get, but it wasn't just about gas. It was about the Bolivian government participating in DEA-led coca eradication and an expression of general angst amid poverty and ethnic tension.

In 2003, strikes and protests led to brutal government crackdowns, and in a pivotal case in September, the military killed six Aymara villagers, among them an eight-year-old girl. The Bolivian Labor Union responded with a general strike that effectively shut the country down. Protestors blocked the highway from El Alto, the slum city of 750,000 mostly indigenous people on a plateau above La Paz, to the capital, choking off La Paz's supply of food and gas. The military responded violently in El Alto, killing sixty-seven people and wounding four hundred.

When Lozada's vice president, Carlos Mesa, broke with the government to protest the use of excessive force in El Alto, Lozada's government began to dissolve. On October 18, recognizing that he had lost control of the situation, Goni resigned and fled the country on a commercial flight to the United States, where he lives now, and from where the new Bolivian government has since tried unsuccessfully to extradite him.

Carlos Mesa took over as president, and in 2005 the government passed a law raising taxes on hydrocarbon exports, but that didn't appease the furious masses, and the protests flared again. In June 2005, Mesa resigned, and a temporary government took power until elections were held the following year. That was the election that brought Evo Morales to power.

Morales was the country's first indigenous leader. Upon taking office, he declared that five hundred years of resistance against colonialism had "not been in vain," and predicted five hundred years of rule by the country's indigenous majority. As part of his inauguration, Morales gave an offering to Pachamama, or Mother Earth, and then in a ceremony at a pre-Incan temple he donned a Sun Priest costume and accepted a gold-encrusted baton intended to signal that the indigenous people were, at last, retaking their country. On May 1, 2006, Morales announced full nationalization of Bolivia's natural gas industry.

Under Morales, relations with the United States deteriorated. Morales is, after all, a coca farmer who is fond of finishing speeches with the cry "Death to the Yankees." Yet Morales's death cry has a surprisingly nonthreatening ring to it. He is no Mahmoud Ahmadinejad. He's not even close to being the most radical indigenous politician in Bolivia. Still, in 2008, the Morales government stopped cooperating with the U.S. Drug Enforcement Agency; that same year Morales expelled the U.S. ambassador to his country, Philip Goldberg, accusing him of supporting separatist movements in the eastern lowlands and saying that "the ambassador of the United States is conspiring against democracy and wants Bolivia to break apart." The next day the United States retaliated, declaring the Bolivian ambassador, Gustavo Guzman, persona non grata.

Around the same time, foreign businesses were lining up to woo Morales for a share of the Uyuni lithium reserves. Mitsubishi, Sumitomo, the Chinese government, and others have all approached the Bolivian government about its lithium reserves.

Perhaps the most aggressive suitor has been the French industrialist Vincent Bolloré. In February 2009, Evo Morales traveled to Paris, where he was greeted by Nicolas Sarkozy, a good friend of Bolloré's. Morales told executives at Bolloré that he expected any foreign company given access to the Salar de Uyuni to seed an electric-car industry in Bolivia. According to the Associated Press, Morales was "surprised" at the positive reaction he got from Bolloré's executives.

Nonetheless, Bolivia still appears determined to develop its lithium reserves on its own, even if it takes longer than it otherwise would, which it will. This is a matter of economic security and sovereignty, goes the Morales message. Enough exploitation. Enough sending raw materials straight out of the country with only a few centavos left behind for the people of this country. What Morales wants is to first develop natural resources inside of Bolivia, in Bolivian-owned plants using Bolivian labor, and then export the value-added products at a fair market price. Bolivia's vice president, Garcia Linera, says this approach to business isn't communism or socialism: it's "Andean capitalism."

The man Morales put in charge of the scientific committee that oversees the Bolivian lithium initiative is Guillermo Roelants du Vivier, a fifty-something Belgian who's lived in Bolivia since 1981. Roelants is a controversial figure in Bolivia. An engineer by training, he says he first came to Bolivia as a volunteer, working with quinoa farmers to improve the quality of their crops and to get them better prices. Within a few years he had formed Tierra, a boric acid company that extracts boron from a salar on a fifteen-thousand-foot plateau near the Chilean border. Tierra was billed as an antipoverty program of sorts—a for-profit company whose profits are all designed to "go back into the community."

In 2003, however, Tierra found itself in the middle of a firestorm. Government agents raided the company and accused it of diverting sulfuric acid—which, because it's used in the production of cocaine, is a controlled substance in Bolivia—to the coca-growing Chapare region in the east. The case drew international attention for its sheer sketchiness. Tierra had been recognized by the United Nations as an exemplary poverty-elimination project. Tierra's employees claimed that the whole operation was the work of foreign companies who wanted to eliminate the competition. Three years after the raid, Roelants and three other Tierra executives were given long prison sentences; Roelants got twelve years, and the government confiscated Tierra. The company's employees descended upon La Paz and protested Roelants's sentence for months on end. Eventually, Roelants was released, Tierra was handed back over, and things gradually returned to normal.

Then in 2008 Morales made Roelants his point man on the lithium project. "President Morales asked me to give him some advice on what to do with Uyuni, because there were not any projects in twenty-five years in Uyuni," Roelants said. "I presented him with a project for a pilot plant for lithium and for potassium, and he asked me immediately to do some design on that plant. Three months later we met again, and the government approved this project."

Now, as a white foreigner in charge of a much-hyped mining project in a country that is extraordinarily sensitive about the treatment of its natural resources, Roelants is a lightning rod. He's been accused of intentionally delaying the lithium project for personal gain, to better exploit concessions he has for lithium and borax in other salars. When articles appear denouncing him, which is often, the online comments on the websites of newspapers like *El Potosí* fill with screeds calling him a drug trafficker (a reference to the Tierra incident) or the Belgian "Rasputin."

One afternoon in La Paz, I met Roelants at a café near his office. A harried-looking, casually dressed white-haired man, Roelants walked into the café and slunk into a booth, practically yelling into his Black-Berry in Spanish about lithium and potassium concentrations in some recent analysis of salmuera. He sounded pleased with the results of the tests, but he couldn't have been having a great day. That morning in *La Razón*, a newspaper generally considered to be friendly to the Morales administration, a columnist lambasted the government for its continuing failure to develop the Salar de Uyuni's mineral reserves. A few pages later another story made it clear that Bolloré was putting serious pressure on the Bolivian government to accept its proposal for exploiting the Uyuni lithium deposits.

This was on top of several months of delays at the pilot plant and the recent collapse of a plan to create a new, quicker-moving agency called Empresa Boliviana de Recursos Evaporíticos, or EBRE, to deal with all evaporative minerals. Just weeks earlier, that plan had fallen victim to parochial politics. It was regional election time, and politicians in Potosí, Uyuni, and elsewhere were upset that the seat of this new agency would be based in La Paz. In traditional Bolivian fashion, local unions threatened to strike if the agency was based in La Paz, and the government responded by putting the plan "on hold," as Roelants put it. Many observers of the lithium project saw the plan's failure as the latest sign of governmental ineptitude. According to the South American business publication *Business News Americas*, "Bolivian President Evo Morales' decision to withdraw Supreme Decree 444, which would have created state evaporitic resources company EBRE and put it in charge of developing the country's lithium industry and other evaporable resources, is an indication that the government still does not have enough experience to handle projects of this size."

Roelants had warned me before our meeting that he could talk only about technical matters—nothing political—but in Bolivia it's difficult to separate the two. After we ordered coffee, I asked him how lithium fits into Bolivia's complicated centuries-long history as a mineral-producing country. He responded with a grimace. "Well, first of all it's better to speak about evaporatives than only lithium," he said. "The Salar de Uyuni has a lot more potassium, magnesium, boron than lithium. Lithium is the new fashion, but economically speaking, for us potassium and potassium chloride is more important than lithium carbonate. Why is that? The price of potassium chloride is about $500 a ton, and lithium chloride is about $6,000 a ton. But the volume of potassium is about twenty-five times higher in Uyuni than lithium. And the cost of making potassium is very low. Through a relatively simple plant you can achieve acceptable commercial quality of potassium chloride. So the cost is low, the price is acceptable, the volume is huge, and the demand is huge—in Brazil, in Venezuela, in Colombia, and of course in other countries." In Roelants's emphasis on potassium, some Bolivians detect ulterior motives: If the future is going to run on lithium, why is the foreigner in charge of developing our world-beating stash of the mineral downplaying its importance? What is the angle here?

Roelants seems fatigued by the constant attention paid to this project, and he is quick to manage expectations. The evaporatives business, lithium and potassium combined, is "huge," he said, "but it will not change the Bolivian state. For instance, hydrocarbons will be for many years more important for Bolivia."

He might be so quick to downplay the project because of the incredible controversy that it, and his involvement in it, has generated.

Two weeks earlier, the villagers who live near the San Cristóbal mines—an enormous silver, lead, and zinc operation, owned by the Japanese trading house Sumitomo, that happens to be a short drive from the government's lithium pilot plant—had staged a good old-fashioned Bolivian resource protest. Primarily they were protesting the mine's exorbitant water usage—more than six hundred liters of water a second, which would go a long way toward hydrating human beings in the high-altitude semidesert. But there was more to it than that. The mine uses too much water, doesn't pay taxes on its water usage, and, according to the protesters, Sumitomo hadn't delivered on its promises to build infrastructure for the citizens of the region. Those citizens responded by seizing eighty loads of ore, shutting down the railroad that connects the mine to the Chilean coast, and threatening to burn down the mine office in the Bolivia-Chile border town of Avaroa. The protest occurred at an embarrassing moment, during the World People's Congress on Climate Change in Cochabamba, a marquee event that drew press and delegates from around the world and provided a bright international spotlight moment for Morales. The government took its time responding to the protesters' demands, in part, some suggested, because one of the protesters' twelve demands was a sensitive one: that Guillermo Roelants be expelled from the country.

Roelants told me that Bolivia will have to play a limited role even in a global lithium market vastly larger than the 120,000 tons of lithium carbonate equivalent shipped each year. "Bolivia will not be able to sell more than 30 percent of the world market," he said. "If they do more than that, they will drop the price down immediately, and you'll have wars on lithium prices, and that's not ideal." The current plan is for the country's industrial-scale plant, which is scheduled to go online in 2014, to produce 30,000 tons per year of lithium carbonate and 700,000 tons of potassium.

Lithium-demand forecasts get hazy starting in 2014. In early 2010, the most current forecast predicted that consumption would increase to 147,000 tons of lithium carbonate equivalent per year in 2013, and that there was "significant potential for increased demand from the mid-2010s as the electric-vehicle rollout gains momentum." In other words,

2014 is the year beyond which no one really has any idea what is going to happen, except that demand is going to increase by some percentage. Everything depends on how quickly lithium-ion-powered electrified cars catch on. If electric vehicles reach a 5 percent adoption rate by 2020, according to the British research firm Roskill, whose statistics industry insiders take the most seriously, existing lithium-production capacity won't be able to keep up with the new demand. Under that scenario, in 2020 the automotive battery market will soak up 60,000 tons of lithium carbonate, up from essentially zero in 2009.

New production wouldn't necessarily require new mines, however, and it doesn't mean that the rise of the electric car is contingent upon the success of the Bolivian lithium initiative. As Keith Evans said at the Lithium Supply and Markets conference in Las Vegas three months earlier, "Large-scale EVs are not dependent on development of the Salar de Uyuni. That is the story that the press has got hold of and repeated and repeated and repeated."

Just as the members of the lithium Oligopoly like to scare off potential competition by boasting about the vast resources they're already sitting on, the Bolivians also like to haul out dubiously large numbers. While the official U.S. Geological Survey estimates say that the Salar de Uyuni contains some nine million tons of lithium carbonate equivalent, Roelants told me that the number is much higher—more like one hundred million tons.

Roelants then acted blasé about such a big number. "[Total] reserves are not so interesting for us," he said. "The salar is so huge and the quantities so big that it's not important if we can verify 50 million tons of reserves or 100 or 120. It's more than enough to work for 100 years or 120 years." And this, Roelants said, is why Bolivia can afford to be particular about how it develops its resource. It basically has enough lithium to last forever. Why rush?

Uyuni is a mud-colored village on a mud-colored plain seven hours south of the nearest sizable city. It is accessible by train or via one of Bolivia's most notoriously uncomfortable bus rides, a bone-shaking overnighter on an unpaved byway that's more like a dirt-bike track than anything

The so-called Lithium Triangle—the arid, high-altitude region where Chile, Bolivia, and Argentina meet—is home to a large number of salt flats that are exceptionally rich with the element.

suitable for passenger transport. The plains surrounding Uyuni are speckled with plastic bottles and bags for miles in every direction. On the village's dusty streets, backpackers on their way to the salar mix with indigenous Bolivian women in their traditional bowler hats and a large contingent of surprisingly healthy-looking dogs, which prowl the town unattended. Amid the altiplano grit, however, is a surprising selection of restaurants and bars catering to travelers bound for the salar, including a place across the street from the train station called the Lithium Club.

Upon my arrival in Uyuni, a message was waiting at my hotel: Francisco Quisbert, head of Frutcas, the Regional Federation of Peasant Workers of the South Altiplano, was looking for me. Roelants had arranged for either Quisbert, his daughter, or his son to drive me to the lithium pilot plant, which is located on the edge of the salar about an hour and a half outside the village of Uyuni. Our bus had been about an hour late, and Quisbert was ready to go. (As if trying to demolish the reputation of Bolivian people as unfailingly late for every appointment, every Boliviano I met was unfailingly punctual.)

We drove out of town and across a brown, trash-strewn plain before ditching the official road for a series of makeshift 4×4 trails, which were much smoother than the brutally corrugated infrastructure. The landscape changed from brown rock and dirt to red rock and dirt and back again. Occasionally a small group of copper-colored vicuñas watched us from the barren flats alongside the road-trails, grazing on what looked like nothing but dust.

After an hour and a half of off-roading, we arrived at a boron-mining camp alongside the river that gives the village of Rio Grande its name, a brackish brown stream that, by sliding beneath the salt crust from the southeast and replenishing the brine, is the greatest steady source of water near this corner of the salar. We cut through the camp, passing man-size heaps of white mineral powder, before turning north and continuing on toward the salar, a miragelike white streak in the distance.

Finally, we approached the *planta pilota*. In May 2008, Evo Morales touched down here in a helicopter for the pilot plant's groundbreaking ceremony. Wearing a ceremonial Aymara costume, he gave a mildly belligerent speech explaining the importance of the project. "Bolivia has the largest resources of lithium in the world," he said. "That is why neoliberal governments and transnational companies sought to seize ownership of these resources."

The plant site sits at the base of a bluff several hundred yards south of the salar. During my visit, it was a near-comatose construction site. At 11:30 a.m. on a Friday, it appeared deserted, save an adolescent-looking military guard, dressed in fatigues and passed out on a bench beside the entrance gate. Our approach roused the guard, and soon two other young soldiers joined him in an exercise of standoffishness. Roelants's phone call to the plant the day before hadn't made much difference. It was obvious that they had no idea who we were or why we were there. Ivan, the man who should have been expecting us, was for some reason back in Uyuni; Quisbert gently pressed the guards for several minutes, and eventually they took our passports, handed us hard hats, and allowed us in.

The pilot plant consisted of the shells of three buildings: a two-story office building, a similarly sized lab, and the center of it all, the plant itself. Or what would one day be a plant. At the time it was a structure of rustic-looking wooden beams, two of its three stories covered with dull red brick. No one was working on the building that day, but Roelants had told me that this would be the case; work on the building had paused for the rainy season and would resume in earnest next week. (Of course, the fact that construction work had to pause for the rain raised questions about the ability to mine lithium year-round.) Quisbert pointed to the top of the bluff, where a Bolivian flag flew next to the multicolored, checkered flag of the Aymara tribe, and we scrambled halfway up the bluff to get an overhead view of the site.

It was impossible to draw any firm conclusions about the likely success or failure of the project simply by looking at these buildings. The buildings weren't really the issue. This small construction project, which was several months behind schedule, would eventually be finished, as long as the money kept flowing in, the Morales government remained in place, and the locals didn't burn the place down in an effort to chase Roelants out of the country. But the real question involved the technology that would go inside the buildings and into the evaporation pools out in the distance, on the salar. Without any technical assistance from companies that specialize in projects like this, could the Bolivian initiative manage to overcome the problems unique to the salar—the high magnesium content of the brine, the fact that a rainy season would dilute the brine when it should be evaporating, and that the salt flat surrounding the evaporation ponds would be covered in water for months each year? Perhaps more important, even assuming the Bolivian initiative solves all these problems by its 2014 deadline, will carmakers and governments and trading houses be interested in buying lithium from the salar? Or in the meantime will they find what they're looking for elsewhere—across the border in Chile or Argentina, or in the clays of Nevada or the pegmatites of Australia?

We gave our hard hats back to the guards and retrieved our passports, and then we drove north to the edge of the salar, where the pilot plant's evaporation pools are located. The sky was broodingly overcast, and it began to rain. As we approached the southern rim of the salt, the mudscape gradually became frosted white. The white thickened, then the mud gave way to salt. We drove on a rock-and-salt road into the salar; two Doric-looking pillars composed of stacked slabs of salt stood at the end of the road, a pointlessly dramatic gateway to a sheet of white that appeared to extend to infinity.

Because of the constant influx of minerals from the Rio Grande, this corner of the salar is the richest in lithium, potassium, and boron. It was also covered in several inches of water, giving us a glimpse of what the whole place looks like during the rainy season, when the salar floods to a depth of about one and a half feet throughout. Shallow pools near the edge of the road glowed a faint chartreuse color, a hint at the myriad minerals dissolved in the puddles. Quisbert signaled to his son to stop the truck, and the two of them hopped out, grabbed a blue nylon tarp from the cargo bay, got down on their backs, and set to shielding the engine from water. Apparently, we were getting ready to drive into what looked like an ocean.

Finished with the engine, Quisbert stood up and pointed to the horizon: "The *piscinas* are out there." We got back into the car and rolled toward the water. As the younger Quisbert guided the vehicle onto the salar, Francisco turned to the backseat with a massive grin on his face. "*Una aventura, no?*"

We eased into the water and drove ahead, as if navigating a hovercraft over a shallow sea. The pillars at the end of the stone-and-salt road receded into the fog, and after a few hundred yards, as the water grew shallower, we approached two cobalt-blue test pools that the Bolivian initiative had cut into the salt. Several minutes later we arrived at a series of bulldozed salt dams, which together walled off what would one day be the pilot plant's evaporation pools. We climbed up onto the salt dam to survey the handiwork of the government project. The pools were large, but they were, at the moment, only walls of salt. Before they could be filled with brine and used for evaporation, they would have to be lined with sheets of PVC plastic. According to the official schedule, next year these would be filled with salmuera, and when that brine was sufficiently concentrated it would be piped back to the completed pilot plant to our south. Although the ponds under construction were impressive, it wasn't at all clear how well they would work during the rainy season, when water would lap at the lip of the wall we were standing on.

After leaving the piscinas, we took the salt road back to Uyuni, heading straight across the salar. We exited the salar in a village called Colchani, which is, truly, the salt mines. Its residents make a living by raking salt from the salar into mounds, and then loading it into trucks that haul it to a processing plant on the eastern edge of town. The other line of business here is making salt into shot glasses and candleholders and statuettes to sell to tourists. Small hotels made of salt stand on the edge of town. That pretty much covers the various ways to make a living from the salt, which explains why the lithium project, the biggest opportunity to arise in this area for decades, has become such an incredible source of hope, frustration, and controversy. The greatest concern is that it will become yet another dashed hope, another opportunity to pull the people out of poverty that, for whatever uniquely Bolivian reason, did not work out the way it was supposed to.

After seeing the world's greatest stash of untouched lithium, I crossed the border into Chile to visit the world's purest and most productive lithium source: the Salar de Atacama.

The corridor between La Paz, Bolivia, and the port city of Arica, Chile, is well paved and well traveled. Down in the Lithium Triangle, however, crossing from Bolivia into Chile is like traveling into the future. Instantly the gravel roads give way to immaculate pavement. Within minutes you descend thousands of feet in elevation, and the brain breathes a sigh of relief.

What you don't see much of, however, is life. This is the Atacama Desert, the driest place on earth. Rain falls on the Atacama desert in millimeters per decade. Some weather stations here have never detected rain, and some riverbeds have been dry for tens of thousands of years. Corners of this desert are so nearly devoid of life that NASA finds them ideal for practice runs of bacteria-detecting Mars robots.

Chileans have put this parched swath of earth to use by mining its minerals since before the arrival of the Spanish. In the late nineteenth and early twentieth centuries, the nitrate mines of the Atacama just about supported the entire country, much as the region's copper mining does today. Oddly, for all its superlative desolation, the Atacama is now a high-end vacation destination. San Pedro de Atacama, the oasis town that is the hub of the era, is like a miniature Santa Fe, New Mexico, a sixteenth-century adobe village with luxurious hotels where rooms cost hundreds of dollars a night. Artisanal jewelry galleries and expensive restaurants and bars dot the just-so downtown area, whose dirt streets have been treated with a salt derivative called *vichufita* for maximum charm with minimum danger of dust storms. Dozens of tour companies offer stargazing tours and horseback trips and sand-boarding expeditions to the crowds of gringo backpackers who clog the town square.

Andrés Yaksic, a marketing manager with the Sociedad de Química y Minera—SQM, or as it is sometimes called, Soquimich—had flown up from Santiago to meet me. SQM is the world's largest supplier of lithium and runs the bigger of the two lithium operations located on the Salar de Atacama. The night before my tour of SQM's operations, which were located about an hour and a half to the south, we went to dinner at a hip open-air bar-restaurant in the center of town. Locals and travelers stood around an open fire pit, drinking beer and cocktails. A few tables away were four men wearing SQM jackets, workers out for a night on the town—an option that wouldn't be as attractive for Rio Grande–based Bolivian lithium workers.

I was thrilled to be sitting at that table. Although only 280 miles separate Uyuni and San Pedro de Atacama—even considering the horrible state of the roads, it should be only about an eight-hour drive between the two towns—there is essentially no way to travel between them except to take one of the popular three-day jeep tours of the Salar de Uyuni and Bolivia's stunning Eduardo Avaroa Andean Flora and Fauna Park. Yet Bolivian tour operators are forbidden from crossing into Chile, and vice versa, so the tour involves a handoff between companies at the barren, almost-three-mile-high mountain pass that serves as the border. When we arrived at the border that morning, no Chilean bus was waiting for us. Our guide suspected that the inch or two of snow that had dusted the mountain roads might have inspired Chile, which is overcautious about bad-weather driving, to close the road leading to the border. But we had no way of knowing; the small immigration office on the Bolivian side of the border had no phone, no radio, no form of communication with Chile. Word spread among the stranded travelers that no bus had come yesterday, and most likely no bus would be coming today; the only two reasonable options were backtracking and hoping to find room in the hostel on the other side of the volcano—and then return to the border every day until finally making it to Chile—or pooling cash to hire a driver to haul us the seven hours back to Uyuni. Finally, a Chilean highway inspector arrived and told us that within a couple of hours a bus should be there. The authorities had decided that the road, which had not a flake of snow on it, was safe for driving.

Andrés asked me about the Bolivian pilot plant, so I described my impression of it in broad strokes. He was curious, not concerned. "Aren't they making it out of bricks?" he asked. They are, actually. Is that strange? "Yes, that's pretty strange," he said.

I understood what he meant the next day when he drove us to SQM's enormous operation in the Salar de Atacama, a dusty-brown, rocky-looking flat that lacked all of the beauty of the Salar de Uyuni but offered the highest lithium concentrations of any known salar on the planet. Here the plants, the actual machinery used for turning heaps of raw material into industrial chemicals, looked like *plants*—refinery-looking productions made of metal pipes and tubes and smokestacks

that hummed and emitted pneumatic *pffts* and hisses, signaling that
industry was indeed at work.

If the Bolivian pilot plant was the lithium-mining equivalent of
a subsistence farm under construction, SQM was an agribusiness giant. Dozens of buildings and trucks and plants and evaporation pools
and hills of white mineral sprawled as far as I could see. Giant rolls of
black plastic, used to line the evaporation ponds, stood alongside the
roadway. Satellite images of this place show a collection of large blue
squares carved into the coffee-colored salar, like the world's greatest
swimming facility, located in the middle of nowhere. This is SQM's
collection of more than one hundred evaporation pools, where the
brine that's pumped up from underneath the salar is left to bake in the
desert sun, concentrating the solution and starting the process of extracting potassium, boron, magnesium, and lithium. From the road
that enters the salar—a firm, smooth surface coated with magnesium
chloride, a waste salt extracted during lithium production that SQM
is now marketing as a product for treating unpaved roads—white
mounds of processed salt stand in the distance.

All of this land, from the salar to the mountains in the east to the
Pacific three hours to the west, once belonged to Bolivia. The War of
the Pacific, sometimes referred to as the Saltpeter War, changed that.
The conflict began in 1879 over rights to mine sodium nitrate, a fertilizer
suddenly in great demand, from the Atacama Desert. Bolivia had discovered that a contract that supposedly allowed the Antofagasta Nitrate and
Railway Company to avoid paying taxes on sodium nitrate (Chile saltpeter) was incomplete. When Bolivia attempted to levy taxes on the company, the company refused, and Bolivia threatened to seize it. Chile
responded by sending five hundred troops to occupy Antofagasta, and the
war began. By the time the fighting ended in 1884, Bolivia had lost its
coastline. Chile gained the territory that is still home to vast supplies of
nitrates, the world's largest copper industry, and the mineral deposits dissolved in the brine a few meters below Yaksic's rented pickup truck.

According to the terms of the 1904 Treaty of Peace and Friendship,
Chile must allow Bolivian commerce to pass through its ports, and a
functioning railway connects Uyuni to the Pacific city of Antofagasta,
passing within five miles of the pilot plant in Rio Grande. But that

doesn't change the intimidating reality that the Bolivian lithium project, once operational, will have to compete with a large corporation that sits nearly three hundred miles closer to the coast, at a far more comfortable altitude of seven thousand feet, in a wealthier and more stable nation. The fact that this Chilean salar and the port city that delivers its bounty to the world once belonged to Bolivia—that's the insulting part.

The insult is still fresh in the Bolivian mind. The Bolivian government still claims a right to the coastline and even maintains a navy, which is stuck patrolling Lake Titicaca. An annual holiday, Día del Mar, commemorates the loss of its access to the sea. A martyred soldier from the war, Eduardo Avaroa, is honored with the eponymous national park and with a statue in the square across the street from parliament in La Paz. Since negotiations between the two countries over renewed access to the sea failed in 1978, Bolivia and Chile have had no diplomatic relations. Among Bolivians there is widespread resentment of and distrust for Chileans, which helps explain why the reaction among the Bolivian public to the natural-gas pipeline through Chile was violent enough that it cost approximately a hundred lives and the jobs of three Bolivian presidents.

We drove to a small management building and gathered in a conference room, where coffee, sandwiches, hard hats, boots, safety vests, and a PowerPoint presentation were waiting for us. Yaksic gave us an overview of the business. SQM has been working in the Salar de Atacama since 1996, when it started pulling potassium from the salar's brine to make fertilizer. That year the company began selling some lithium (which it was already extracting in the process of producing potassium), but it didn't expand its lithium business until 2008, when the market began to warm up. Even now, however, lithium is a small portion of SQM's total business; they supply 31 percent of the world's lithium, yet that provides only 8 percent of the company's revenues. Their main market is "specialty plant nutrition," a business in which they command as much as 50 percent of the world market share. SQM's nitrogen-based fertilizers still come from saltpeter, or caliche. SQM supplies a quarter of the world's iodine, which it also extracts from caliche. But the caliche sources are to the north of here; in the salar, it's all about salt and the potassium, boron, and lithium extracted from it.

The saga of SQM's effective owner was not part of Yaksic's Power-Point presentation. The head of SQM's biggest shareholder, Pampa Calichera, is a man named Julio César Ponce Lerou. Ponce has been accused of being the beneficiary of blatant nepotism. A forestry engineer, Ponce married General Augusto Pinochet's daughter Veronica four years before the dictator took power in a 1973 coup. A year into Pinochet's reign, Ponce returned from Panama, where he had been living, and rocketed to success, holding jobs at the head of a succession of large state-owned companies. He was appointed the director of Corfo, the agency in charge of selling off state-owned assets, and along the way he acquired a massive herd of cattle and a four-thousand-acre estate. In 1988, he led the group of investors that bought SQM from the government. After Pinochet, a government investigation found that the state sold parts of SQM for less than a third of what they were worth. In what might just be a case of purple prose, a Chilean newspaper compared Ponce to Kaiser Soze, the vanishing villain of the film *The Usual Suspects*, a man so secretive and skilled that he almost convinced investigators that he didn't exist.

The presentation did explain what is so special about the Salar de Atacama. Geologically speaking, it is a basin of a little more than a thousand square miles, formed by the movement of the Nazca Plate beneath the South American Plate—the same geological phenomenon responsible for the earthquake that devastated southern Chile a few months before my visit, and that also caused the strongest such event ever recorded, the Valdivia earthquake of 1960, which measured 9.5 on the Richter scale. As in the Salar de Uyuni, freshwater flowing down from the mountains through mineral-rich volcanic rock is what puts the lithium underneath the salar's surface. SQM says that because of this geological accident, there are 40 million tons of lithium carbonate equivalent *reserves* here; that's the measured, economically extractable part. The unrealistically optimistic number is 190 million tons as resource. SQM's current capacity here is 40,000 metric tons of lithium carbonate equivalent per year.

The lithium concentration in the Salar de Atacama averages 2,700 parts per million. That's one natural benefit of the Atacama. The other is that this desert hellscape is the perfect environment for extracting evaporative minerals. The quasi-Martian sun toasts the brine in these evaporation ponds more harshly than at any other lithium-bearing location on the planet. Water evaporates at a rate of 3,500 millimeters per year here; the next highest evaporation rate in the lithium industry is found on the Argentine Puna, several hours to the southeast, where water evaporates at a rate of 2,600 millimeters per year. In Uyuni the rate is 1,300 to 1,700 millimeters per year. Unlike the Salar de Uyuni, the brine of the Salar de Atacama also has a very low magnesium-to-lithium ratio, which makes processing Chilean brine into the finished lithium product easier and cheaper than brine from the Salar de Uyuni. All these benefits, combined with the existing infrastructure (much of which was put here for the copper-mining industry), make SQM comfortable boasting that, if they really wanted to, they could deliver three to four times the total global demand for lithium.

After the presentation, we met Álvaro Cisternas, a manager of harvesting operations. In his pickup truck we drove out among the vast field of evaporation pools. An autumn breeze made for a comfortable day, but the sky was the same as it always is in the Atacama: blazingly clear. The sunlight felt unfiltered, like a warm, tangible fluid that puts active pressure on the skin.

We stopped at a small wellhead, which draws brine from 120 feet below the salt through a tube about the girth of a firehose. But it was hard to pay attention to anything except the evaporation pools all around us, massive sheets of cerulean water bordered by salt-white banks. For an industrial-scale chemical-processing tool, the evaporation pools were gorgeous. As the water evaporates, the salts that are dissolved in the brine precipitate out in an orderly series. First goes the sodium chloride, or halite (table salt), which solidifies and settles to the bottom of the pool. The brine is transferred through a series of pools until it's concentrated enough that the next essential salt falls out: potassium chloride, or potash, a key ingredient in the plant-food pellets we feed our hydrangeas. After that a mineral called carnallite, a salt of magnesium and potassium, begins to settle. Next comes bischofite, the magnesium salt that coats the roads leading through the salar.

In the majority of these pools, the process is stopped when the desired potassium-based salt precipitates out in good measure. Then the pool is drained, leaving behind a mass of wet, blinding-white salt. In one such pool an employee in a blaze-orange vest walked about on a field of salt, measuring the remaining water levels. This potassium-based salt would be fed into an on-site potassium chloride plant, which would transform it into market-ready fertilizer.

Before long we began to drive past a series of pools of deepening shades of chartreuse. These were the lithium pools. As the brine reaches ever-higher concentrations of lithium, it yellows, eventually reaching the glowing orange color of Tang. (The magnesium and the lithium provide the yellow hue.) The final product, a solution of 6 percent lithium, is a sickly yellow-green that probably resembles the urine of someone who's been stranded for a couple of days in the middle of the Salar de Atacama. It takes roughly fourteen months of being transferred from one evaporation pond to another for the lithium concentration to rise from 0.2 percent to 6 percent; any higher and the lithium begins to precipitate.

At the end of this series of evaporation pools, tanker trucks pull up to the *plataforma despacho litio*, fill up on the yellow brine, and then haul it to SQM's lithium carbonate plant, three hours away on the Salar de Carmen, just outside Antofagasta. There it's processed into lithium carbonate, a white powder that looks so much like cocaine that I didn't dare try to fly back to the United States with samples.

While most of the mounds of salt scattered throughout the facility are made of feedstock for further refinement, one of them is set aside for the purpose of gawking at SQM's sprawling salar operation from above. To conclude our tour, Cisternas drove us to the top of the hill, parked at an overlook, and proudly urged us to take in the view. Evaporation pools, tractors, trucks, outbuildings, and hills of valuable salt stretched for what appeared to be miles, although the air there was so dry and clear and the view so staggeringly uninterrupted that getting a firm perspective on the operation's size was difficult; for instance, the mountain range seventy miles away that marks the edge of Bolivia looked close enough to jog to.

The contrast between SQM and the three-building Bolivian pilot plant was so great that it's almost not fair to compare them, but Guillermo Roelants all but challenged me to do so. As our meeting the week before in the La Paz café came to an end, I mentioned that my Lithium Triangle tour would involve a visit to SQM in the Salar de Atacama. "Ah, good," he said. "Then you will see where we will be in four or five years." Standing at the SQM salt-mountain lookout, I realized that Roelants was being outlandishly optimistic.

After a day of rest in Chile and a twenty-hour series of bus rides back to La Paz, I sat down in a café on La Paz's tree-lined El Prado for a conversation with the Bolivian writer Juan Carlos Zuleta, who has earned a reputation for his close and critical coverage of the Bolivian lithium initiative. In the months before our meeting, he had turned against Guillermo Roelants, whom he suspected of intentionally delaying the Bolivian initiative for personal gain. The government had "no strategy," he said, which he felt was a tragedy. Zuleta fiercely believes that lithium is essential to the future of the Bolivian state.

I had first met Zuleta at the Lithium Supply and Markets conference in Las Vegas, where he closed the conference with a talk that was highly skeptical of the Bolivian government's approach. There were at least three challenges to the Bolivian initiative, Zuleta explained. The first was political. The second involved the array of logistical problems unique to Uyuni—the low evaporation rates, the high magnesium-to-lithium ratio, the lack of easy access to the sea. The last one was social. The local communities were demanding that the government do something about the region's endemic poverty as a condition for harvesting the lithium that lies on their land. So far, the government wasn't meeting those demands, and Zuleta's comments that January turned out to be prescient when the communities around Uyuni erupted in protest and paranoia.

Now, a little over three months after I heard him speak in Las Vegas, Zuleta explained why he had taken such a critical turn in his thinking. Just before Zuleta left for Las Vegas, he said, Roelants gave an interview to a Bolivian business publication in which he spoke of prioritizing potassium over lithium—basically the same approach that Roelants had explained to me the previous week. To Zuleta, a true believer in lithium, this was a "gross error."

Then Zuleta began thinking that it must be something more than an error. "I started to search for answers," he said. "I thought there was something else." Zuleta soon found out that Roelants also had a concession at a salar called Pastos Grandes, which Zuleta believes may in fact contain the highest lithium concentrations in the world. How did Roelants acquire such a concession? Through Tierra. And wouldn't it be convenient for Roelants if the Tierra project began producing lithium before the perpetually delayed Uyuni project did? Or if the Bolivian initiative prioritized potassium so heavily that it left a market opening for the lithium that Tierra would perhaps soon mine from Pastos Grandes?

Zuleta's theory was speculative, but it was a clearheaded version of the thinking of many of the people who live in the region around Uyuni. Pieces published in *El Potosí* and *El Diario* that April make it clear that plenty of the villagers in the department of Potosí see Roelants and Tierra as the latest in foreign interests that are out to steal Bolivia's mineral riches from the people.

The most generous view Zuleta holds is that Roelants and the government simply don't know what they're doing. The consequence of this ineptitude, he believes, could be Bolivia losing its chance to enter the market of the future. "I know that lithium is a big thing—not just for Bolivia, but for the whole world," he said. "Otherwise, how do you explain that all the automakers in the world are betting the next twenty to thirty years on this?"

The night before I left La Paz, I ran into Oscar Ballivian. We hadn't yet been able to get together there, but the week before I had seen his picture in *La Razón*, above the story about the pressure that Bolloré was putting on the Bolivian government. The Bolloré proposal included plans for a Bolivia-Bolloré consortium that would build lithium-ion batteries and, eventually, even electric cars in Bolivia. Bolloré had been courting the Morales government hard for well over a year. The company was clearly tired of waiting for an answer.

As the geologist involved with the lithium deposits of Uyuni since the beginning, Ballivian seems to desperately hope that some sort of agreement can be reached, that the salar's riches can finally be tapped. He feels that the salar was *meant* to produce the lithium that would help transform technology in the new century. "The salar is my life," Ballivian said.

The next morning, Ballivian was leaving for Argentina, where Bolloré was drilling for lithium in a small salar near the city of Salta.

THE GOAL

In the quest to rid our cars of oil and our grid of coal and gas, battery scientists have at least two essential duties. The first is to continue to grind through the periodic table in search of the incremental advances that will steadily make the technology a little better every year. The second is to chase ideas that may be decades from commercial reality, because while everyone else is arguing about state tax credits for pack assembly plants and the price of separator material, somebody has to.

Three and a half decades after Exxon built and then killed his breakthrough rechargeable lithium battery, Michael Stanley Whittingham is still at it. Since 1988 he's been a professor at the State University of New York at Binghamton, on the banks of the meandering upper Susquehanna River. After a detour with another oil company and then a few years spent on (what else?) high-temperature superconductors, Whittingham has been back at battery research since the funding returned in the early 1990s.

Whittingham's domain, the Institute for Materials Research, is located in a faux-Bauhaus classroom building in the middle of the campus. When I visited one autumn day, we sat in his cluttered office and talked about his time in the field. Early in our conversation he smiled, reached over to his desk, and handed me a block of clear plastic. Inside it was embedded a vintage Exxon lithium titanium disulfide battery and a digital clock display. The clock wasn't accurate, but after more than thirty years, it was still running.

I asked him what major challenges battery scientists had to overcome. "Everything, almost," he said. "How can you make new materials cheaper? Then you need a good electrolyte. You need a better separator than they've got right now. And you need totally different materials for them. One area of interest is better geometries for cells . . ." His response reminded me of something Jon Lauckner once said: "The ideal goal will be to have the same energy density as gasoline or diesel fuel. That's where we'd say, okay, we've arrived."

Whittingham's lab, in the basement below his office, focuses on electrodes, on finding the revolutionary material that will push things forward in a significant way. In places, the lab looks impressively high-tech, sanitized, expensive. In others, it looks like a tire shop. Workbenches are topped with brutal iron devices that pound powders into pellets suitable for annihilation in a furnace, where Whittingham's students study the composition of materials by torching them and taking detailed measurements of their oblivion.

Searching for new electrode materials is a matter of first synthesizing the raw active material you want to test and then building tiny batteries in which you can try it out. The process starts with measuring spoonfuls of chemicals out of hand-labeled canisters, as if making a cake for some arsenic-based life-form from another planet. This recipe is then baked into the active ingredient under examination.

To get a portrait of the inner structure of the molecules under study, battery scientists use X-ray diffractometers, which determine the molecular shape of a material by bombarding it with rays and analyzing the pattern of the ones that bounce off. These rays carry information about the atoms that reflected them—the number of atoms they're bonded to, the length and angle of those bonds, the behavior of the electrons flitting around them. In Whittingham's graduate school days, you had to take the measurements from the XRD and piece together a picture of these structures by hand. "Something like that"—in his office he pointed to a waist-high ball-and-dowel model—"would have been an entire Ph.D.

thesis." These days, computers do the work, but it still takes days even for a machine to "solve" the structure of a single crystal.

The test batteries that use these materials are built by hand using a tabletop version of the process employed in the battery factories. Mix a couple of teaspoons of the powdered battery material with binder, an equal measure of carbon black, and a solvent. Paint that concoction onto a sheet of aluminum foil. Bake it in a miniature curing oven. Transfer it to a vacuum chamber that extracts every trace of moisture. Remove the film of cathode material and use an industrial-grade hole punch to slice out a quarter-size circle of pencil-lead-gray electrode material.

Next, the battery cells are assembled in a glove box, which looks like a giant aquarium with inside-out plastic gloves extending like udders from its front. The glove box is filled with either helium or argon, inert gases that won't react with the pure lithium metal inside. Hands inserted into those udder-gloves, a scientist has at his or her access all the ingredients of a small experimental battery—a coin cell, as it's called, because the finished product looks like a blank half-dollar. First, using a pressurized air lock, you slide the cathode material into the helium chamber. Two casing discs, which each look like the half shell of a pocket watch, sandwich everything that goes in between. Take one disc. Lay the freshly baked cathode inside. Take the other disc. Squeeze in a circular gasket, which will keep the cell airtight once it's finished. Slice a little lithium off the roll, and using a small die, cut a circle of metallic lithium, as if pressing a cookie cutter into rolled dough. Put the disc of lithium metal—the anode—into the left disc. Reach into the bag of circular ceramic separators and lay one on top of the anode. Take a dropper and soak the separator with five or six drops of electrolyte. Then drop in another gasket, squeeze the sides together, and slide the coin cell into a vise that presses it together. One small coin-cell battery, ready for testing.

Upstairs in the cycling lab, two wooden shelves are lined with these electrical half-dollars, each one wired to a power source and a computer. The idea is to charge and discharge each cell repeatedly while monitoring its behavior—how much energy it can store, how quickly it can dump that energy, how many cycles it can stand before beginning to fall apart, and so on.

The insights gained from these studies allow scientists like Whittingham to puzzle out what, exactly, is happening inside these materials. Are ions traveling into the crystals and bonding where they're expected to? Are they coming out when we coax the reaction into reverse? Is the structure breaking down as a result? This is the kind of work that, far downstream, determines comparatively proletarian things like zero-to-sixty times and highway passing power.

And those are the things that get people like MIT's Donald Sadoway excited about electrochemistry. "After you've driven an electric car, you don't want to go back to internal combustion," he said, "because all electric drive is just neck-snapping acceleration and flat torque curve—zero to thirty is neck snapping, fifty to seventy is neck snapping. It's an exquisite ride. I've said for as long as I can remember that the problem with environmentalism in this country is that it's been largely in the hands of the crunchies, and the mainstream American views it as sort of a penance, or a duty. People fail to realize that it's a chance for a new start, a chance for reinvention and a way of making things—not just, 'Well, we'll compensate and we'll try to make things as good as we have them now but with lower carbon intensity.' It's a chance to make things way better!"

This line of conversation sent him into full pontification mode. "I teach this big freshman chemistry class here, and I do a little bit of editorializing from time to time, and I tell them to challenge themselves. Instead of thinking about a process to make steel that does minimal harm to the environment in comparison to the process we have right now, what about thinking about a process that actually *cleans* the air, and *cleans* the water, so that people fight to have the smelter sited *in* their neighborhoods because the trees are greener near the smelter, the water is cleaner having passed through the smelter? A car that drives down the road and its exhaust is purer than the air coming into the front of the car? Why aren't you thinking this way? Why are you thinking that the best we can do is zero? Today we're at a negative, and the best we can hope to achieve is to get as close to zero but on the negative side as possible? And I'm saying, why can't you bust through the zero axis and go positive? Why not? It's only the limits of your imagination that prevent it."

It's impressive that longtime battery scientists, people who witnessed the stillbirth of electric-car revivals in the 1970s and the 1990s, can still muster that kind of imagination. Consider the case of Elton Cairns.

Cairns led General Motors's research into high-temperature electric-vehicle batteries in the late 1960s. He now holds posts at the University of California at Berkeley and Lawrence Berkeley Laboratory, where he specializes in the kind of next-generation lithium-battery technology that could go a long way toward meeting Lauckner's goal of matching the performance of the gas-powered engine: lithium sulfur.

Conceptually, the battery is simple. Composed of a lithium-metal anode, an elemental sulfur cathode, and an electrolyte made of a mixture of ionic liquids, liquid polymers, and a lithium salt, it works on the reaction of lithium and sulfur to form Li_2S. Theoretically, the promise is immense: a lithium-sulfur battery should be able to store five times the energy of the lithium-ion batteries we use today.

One of the immediate benefits is getting rid of carbon. The anodes in all of today's lithium-ion batteries use LiC_6, which by weight is only 10 percent lithium. Replacing all that carbon with active, charge-carrying lithium would be an instant advantage.

Sulfur brings significant benefits as well. Because a sulfur atom is about a third of the weight of the cobalt oxide molecules found in today's consumer-electronics batteries, the weight savings is almost as dramatic on the cathode side, particularly since one sulfur atom can bond to two lithium atoms. If a lithium-sulfur battery could be made to work correctly, it could store hundreds of watt-hours per kilogram, enough to jump up into the realm of the several-hundred-mile electric car. The big, fundamental problem is that sulfur is terrible at conducting electricity. "A lot of nanoscale engineering has to be done to make the sulfur electrode work well," Cairns said.

Sulfur-based lithium batteries may be years away, but there's a near-term alternative that moves the technology in the same direction: silicon. Like sulfur, silicon can bond to more lithium atoms than can the materials currently used in lithium-ion electrodes. And lithium-ion batteries that use silicon are very close to production—like the cell that Panasonic will start selling in 2013, which will use a silicon alloy anode to reach an energy capacity of 4 amp-hours, a 30 percent improvement over the highest-energy cells available today.

But silicon could make even greater leaps possible. In 2008, a young Stanford professor named Yi Cui attracted major attention when he unveiled silicon "nanowires" that could replace the carbon in the anode. "One silicon atom can bond with 4.4 lithium ions," Cui said, whereas it takes 6 carbon atoms to bond with one lithium ion. Silicon is heavier than carbon, but even so, on balance, silicon can hold ten times more lithium than an equivalent amount of carbon.

The catch is that when silicon reacts with lithium, it undergoes a huge change in volume, swelling when charged and deflating when discharged. Over time, this strains the electrode and causes it to break down. Cui believes his nanowires work around this problem by shrinking to sizes where mechanical strain is no longer a problem. "If the object is already smaller than the smallest thing you can break," Cui said, "they don't break anymore."

Cui, who in 2008 cofounded the company Amprius to commercialize silicon nanowires, is a sort of battery savant. In addition to his silicon research, he's behind some of the oddest and most creative battery research happening—schemes to make lithium-ion batteries out of things like paper and fabric. Cui has shown, for example, that paper can act as a substrate for a lithium-ion battery. Basically, you take a piece of printer paper, dip it in an "ink" of carbon nanotubes or nanowires, and the paper becomes astonishingly conductive. "Paper is very light, and its internal structure is made of cellulose fibers, so it soaks [the active battery material] up like ink. Once you put the materials in, they can be accessed by the electrolyte very fast for high power." He envisions one day using a factory akin to a paper plant to manufacture batteries.

Donald Sadoway has a different vision, one that gives a glimpse of how the battery could evolve without lithium. A few years ago, he started thinking about the challenge of grid storage and the main limitations of batteries in that application. "Batteries intrinsically store charge, but they don't like high current," he said. "So the gambit for everybody else is, How do we engineer a battery that can handle high current?" Hooked to the grid, they'll need it—to quickly soak up a load of electricity from a fast-running windmill, to quickly dump it into the system when the wind dies down and everyone in the neighborhood turns up the air conditioner.

"What I did was turn the problem around. Why don't I start with a device that intrinsically handles high current and then teach it how to store charge? I know what likes high current: an aluminum smelter. Now, how can I turn this into a battery? What is it about an aluminum cell that fails to make it a two-way street?"

A student of Sadoway's, David Bradwell, did a study of the concept for his master's thesis. "We ultimately came to the realization that we needed liquid metals at both electrodes," Sadoway said. "And that's impossible. Because you make metal at the cathode and you make nonmetal at the anode. That's where the invention came into play."

He grabbed a laminated, place-mat-size periodic table and laid it on the coffee table in front of us. "Over here you have the metals," he said, pointing in the general vicinity of iron, "and 75 percent of the periodic table is metals. And over here are the nonmetals—the fluorine, the chlorine, the oxygen, the nitrogen. But it's not a sharp departure from metal to the nonmetal zone. Along this staircase we have what we call the semimetals, or the metalloids."

He shifted from chemistry-lesson mode to recounting his lightbulb moment. "So I was sitting looking at the periodic table one day and I was thinking, you know, antimony's over here, and it's got an electronegativity of 2.05, which is sort of in the same vicinity as sulfur. And sulfur's a nonmetal. But antimony's a good metal. If you see a block of metal, it's shiny and it conducts electrons. But if you take magnesium, which is way over here"—he points—"it's got electronegativity of 1.3. It's such a good electron donor that if you put magnesium in the presence of antimony, antimony will be intimidated into becoming an electron acceptor. And that was when the lights went on." He had found a loophole.

He realized that antimony and magnesium have similar melting points but vast differences in density, which means that if you melt them both and put them in a bucket, they'll separate. "So now I've got liquid metal, which is insoluble in a liquid salt, which is insoluble in a liquid metal. I've got the three layers. They phase separate. They stratify by density. And I don't need any solid separator! It's salad oil and water."

Aside from the general intriguing weirdness of building a battery out of a layer cake of self-separating molten metals, the great thing about this design, Sadoway said, is that it should, in theory, be eminently scalable. "If you want to scale a sodium-sulfur battery to the size of a thirty-three-gallon garbage can, you can't," because in order for a sodium-sulfur battery to work, the solid beta-alumina electrolyte—the same thing that triggered the rise of modern battery science in 1967—has to be paper thin. Small sheets of beta-alumina can be paper thin, but large ones can't, because they'll break and cause the whole system to fail. This liquid metal design, however, didn't require any delicate paper-thin parts. These liquids should simply fall into place, no matter how many gallons of them are involved. "If I want to make something the size of a thirty-three-gallon trash barrel, I just make something the size of a thirty-three-gallon trash barrel. I've got liquid metal on the bottom, molten salt, and liquid metal on the top. And current leads in from the top and out the bottom, and away you go." That, at least, is the vision. "Right now we're at the size of a little coffee cup."

In 2009, Sadoway received a $7 million grant from ARPA-E, the agency in the Department of Energy that funds high-risk, high-reward research, to determine whether his vision for a battery can work at a large scale. He said that he doesn't care about getting money or fame or a career boost out of the project. In his view, grid storage is an environmental necessity, but the market that any grid storage device will have to compete in is brutal.

"Your competitor is not another battery" when it comes to grid storage, he said. "Your competitor is a gas-fired peaking unit. When the sun isn't shining and the wind isn't blowing, guess where your electrons are coming from? A gas-fired peaking unit. It's a combustor. And I'm saying that instead, if we had a battery, we would have this beautiful situation in which you could draw electrons from the sun even when the sun isn't shining. And that's really compelling. That gets people out of bed in the morning."

What gets quite a few scientists out of bed each morning is the highest-risk, highest-reward battery technology of all: lithium-air. At the moment, lithium-air appears to be the best chance battery scientists have to beat gasoline. It is elegant in concept and, theoretically at least, extravagantly energetic.

For the sake of comparison, consider that a lead-acid battery can store something like 40 watt-hours of energy per kilogram. Today's best lithium-ion batteries can hold about 200 watt-hours per kilogram, and lithium-ion has a theoretical maximum of 400 watt-hours per kilogram. Lithium-air has a theoretical maximum of 11,000 watt-hours per kilogram. Even after handicapping to take into account weight, efficiency, and other foreseeable technological obstacles—after assuming that, for the sake of argument, the lithium-air battery will be able to deliver only 15 percent of its theoretical energy capacity—it still matches what gasoline, because of the terrible efficiency of the internal combustion engine, can deliver. And that is why scientists have been dreaming about it for decades. "As with all things in life where there's a big prize, it's not an easy one to reach," Peter Bruce said.

Lithium-air is probably the purest and earthiest battery chemistry possible, because in its simplest formulation it involves nothing but lithium, oxygen, and carbon—the lightest metal in the universe and two essential elements of all living beings. "You take the positive electrode of a lithium-ion battery and you replace it with porous carbon," explained Bruce, who today is one of the world's leading lithium-air researchers. "The electrolyte"—this could either be an organic solvent as in today's lithium-ion batteries, a combination of polymers, or maybe even something based on water—"floods the pores of the carbon. Oxygen comes in from the air." And then the lithium ions, the oxygen, and the electrons routing around through the external circuit all combine to form lithium peroxide (Li_2O_2), a solid. Then, as with any rechargeable battery, the whole thing happens in reverse. "When you charge up the battery, you actually decompose this solid material. It goes back to lithium ions and electrons and pumps oxygen into the atmosphere again."

Mainly because of the signal it sends to the world—IBM is interested!—the highest-profile lithium-air project right now is Battery 500, a lab dedicated strictly to lithium-air research at IBM's Almaden Research Center. "A practical electric car will need a lot more mileage than is possible with lithium-ion batteries," said Winfried Wilcke, head of the project. "Five hundred miles is the target you really want." That, along with a nice resonance with the Daytona 500, is why IBM decided to call its lithium-air project Battery 500—"to differentiate this from incremental improvements of lithium ion."

The IBM project is taking a supercomputer-driven, fundamental-physics approach to the problem. "Electrochemistry has had a long history of a very Edisonian approach," Wilcke said. "But for something as risky and difficult as a lithium-air project, that's not good enough." It's risky and difficult because "wherever you look there are challenges," he said. "It's like climbing Mount Everest."

First there's the maddening difficulty of recharging. Getting the discharge reaction to happen once, to get the lithium to react with oxygen to form solid lithium peroxide—thanks to recent advances, that part is not a problem. What is a problem is getting that reaction to happen in reverse, to get the solid lithium peroxide to decompose into oxygen and then plate pure lithium back on the negative electrode with mirrorlike smoothness, rather than coating it with the fuzzy metallic spikes that have long been the curse of lithium-metal electrodes. Power is another problem. The reaction between oxygen and lithium is intrinsically slow, far too sluggish to blast a car up the highway in a passing maneuver.

Hope for the power problem comes in the form of nanotechnology, which, as it does for many other lithium-based battery technologies, increases the surface area of the individual electrode particles, thereby dramatically increasing the rate at which the battery can charge and discharge. (Catalysts can also help that reaction happen more quickly.)

As with anything involving metallic lithium, safety is a concern, although Wilcke argues that we should first find out whether lithium-air batteries are even remotely feasible before worrying about safety issues. "A lot of people like to talk about dendrites and anode and danger and so on," he said. "I kind of think this is a secondary project. For one, there is no need really to use metallic lithium in a lithium-air battery. One could use a carbon intercalation anode or something like Yi Cui's silicon battery. You could combine that with an air cathode, maybe get up to 1,000 watt-hours per kilogram, but not to 1,700."

Jeff Dahn, the battery field's most experienced spokesperson on the dangers of lithium metal, is still cautious. "What Moli Energy found back in the late 1980s was that lithium-metal electrodes, just under normal use, led to cell failures that were at just too high an incidence rate to make it a viable business," he said. Nonetheless, Dahn urged me to talk to Steve Visco, who as chief technology officer of the Berkeley company PolyPlus is in charge of making lithium metal safe and usable.

PolyPlus was spun out of Lawrence Berkeley National Laboratory in 1990 as a sort of think tank for lithium-sulfur research. "In many ways it operated as a kind of innovation center for batteries," Visco said. He told me that the company did "all of the groundbreaking work in lithium-sulfur chemistry." As they were studying lithium-sulfur batteries, they found that they couldn't find a way to stop the sulfur from interacting, undesirably, with the lithium. "There was only one real way we could see to stop that, and that would be to somehow encapsulate the lithium with a conductive solid electrolyte, like a thin glass layer."

After doing some basic research, they started looking for an existing material they could use for that protective layer. They were lucky. A company in Japan called Ohara was making exactly what they needed. "I called them and had them ship us some plates, and when I talked to their representative, he said, 'Well, I have some of these plates, but they've been sitting on my desk for a couple years.'" One of the major challenges with fabricating a material like this is making it stable enough to sit on a desktop without reacting with the moisture in the air and corroding. "And I thought, 'Wow, if they're that stable, that they can sit on his desk for two years without turning into a puddle, I want to look at those.' So I said, 'Send me the samples that have been sitting on your desk for two years.' He did, and we immediately put lithium up against those plates after actually verifying it was conductive, and it degraded. So we said, 'Okay, that's why nobody's using it—it's not stable against lithium.'"

Fortuitously, PolyPlus had already developed a process for coating lithium with multiple layers of different materials that were designed to make otherwise unstable combinations of materials—combinations that normally react and corrode or melt or catch fire—stable. "Instead of lithium touching that white ceramic piece, we put something between the two that allowed lithium to move between the two but where nothing would react. We tried that and what we saw was, 'Wow, this looks really stable.' And then at that moment one of our electrochemists and I started talking and we said, 'You know, if this stuff is stable in air, we might be able to build a lithium-air battery.'"

People had been talking about lithium-air for decades, but no one had ever figured out how to get around the fact that air contains moisture, and moisture attacks lithium. "All these discussions about lithium-air batteries, although interesting, had that basic flaw," Visco said. "That if you were to build something, it would be a bit of a novelty. You'd never have anything practical."

To see if they *did* have something practical, they decided to subject their coated lithium metal to the most direct test possible. "We put water right up against it," Visco said. "And we said, 'Either it's going to get attacked and fall apart, or maybe we'll see something.' And it actually shocked us. What we saw was extreme, very stable electrode potential. So then we said, 'Let's see if we can move lithium in and out,' and it just worked like a charm. So we said, 'Wow, this is a big thing.'" That year, 2003, PolyPlus went into stealth mode, and Visco and his colleagues spent the year writing patents.

When they came out of hiding, they began talking about some of the most interesting far-horizon developments the battery world had seen in a long time. Naturally, they applied for funding from DARPA, the Pentagon's advanced research agency, and got it. They began working on two different lines of research. First was lithium-seawater batteries, which could be used to power oceanographic research vessels or military craft. The second was lithium-air. Within the lithium-air program they began studying both primary (one-use) batteries, which today Visco says are working very well and delivering charge capacities of 800 milliamp-hours per gram, as well as the real prize, rechargeables.

PolyPlus's lithium-air battery is an interesting tweak of traditional lithium-ion design. The negative electrode is made of metallic lithium, and the positive is air. Between the two is a ceramic barrier. "In our battery, things are switched around a bit," Visco says. "It almost looks like a piece of glass, but it's white." But the metallic lithium anode is encased in a series of ceramic barriers that allow it to engage in the right reactions while keeping it completely isolated from moisture. "You can hold it in your hand, you can put it in a glass of water, and it'll just sit there," Visco said. "It's completely stable. And as soon as you hook it up to a wire, it becomes active."

To show exactly how stable the coated lithium electrode is, Visco's team built a lithium-water battery in which the water "electrode" is an aquarium inhabited by clown fish. The water in which those fish live acts as the positive electrode for the battery, which is connected to a green 3-volt LED. "In a sense they're swimming inside a lithium battery, and they're completely unperturbed."

As always, there are hurdles to clear. Visco's team has the same problem as all lithium-air researchers, which is recharging—getting solid lithium peroxide to break back down to its constituent parts in an orderly fashion. And they have lithium metal to deal with. The way PolyPlus encapsulates their lithium-metal electrode makes it easy to handle, but that doesn't mean it will be easy to recharge. "No one has ever really shown [rechargeable lithium metal] to be doable yet," Visco said. And that, in part, is why it'll be a long time before PolyPlus's lithium-air batteries are driving our cars. "Even if we commercialize a lithium-air battery, it's going to take a long time before you see battery packs that are large enough and proven and tested enough that you would start thinking about transportation," Visco said.

Today electric cars come with too many caveats. Unless it has a backup gas engine, an electric vehicle will have to be a second car. Only when cities have installed charging stations in every parking meter and every parking garage will electrics truly be practical. Even then, it'll take a nationwide chain of high-power, fast-charging stations or battery-swapping businesses—whatever sort of Jiffy Lube grows out of projects like Shai Agassi's battery-swap company Better Place—before you can take a road trip in the thing.

There are problems with waiting on the infrastructure, however. Consider fast charging, which would allow electric-car drivers to dump their batteries full of electrons in a matter of minutes, making a recharge only slightly more time-consuming than a visit to the gas station. The math isn't promising for the prospect of a major network of electron filling stations. "Let's say you've got a battery that holds 25 kilowatt-hours," Elton Cairns said. "If you want to charge that in fifteen minutes, then you've got to have a 100-kilowatt substation. If you've got something like the Tesla with over 50 kilowatt-hours and you want to charge that in fifteen minutes, you're talking 200 kilowatts. Your house takes 1 kilowatt. If you want to have something like a gasoline fuel station that is all electrical, you're talking about multimegawatts of power at that station. And I just don't see that happening."

There are a couple of ways to react to this sort of discouraging calculus. One is defeatism. The other is research. "Infrastructure gains are the hardest there are," IBM's Wilcke said, "which is one reason [hydrogen] fuel cells haven't worked." That is exactly why Wilcke is now devoting his career to trying to find out whether the lithium-air battery can be made reality. With a breakthrough battery that can deliver a car five hundred miles on a single charge, only the most speed-addled road tripper would need fast charging or battery swapping. Everyone else will charge curbside at the hotel and then get back on the road the next morning. "I'd rather tackle a really difficult technical problem," Wilcke said. "It's confined to being a technical problem, and you don't need a zillion dollars' infrastructure."

"Society needs higher-energy-density solutions," Peter Bruce said. "There aren't many options on the table. We have to explore the options that we have. Lithium-ion batteries will be with us for many years to come, and they'll be key technologies in vehicles." The reason Bruce and others like him have hope for the prospects of a livable, comfortable postoil civilization powered by electrons snared from the sun and generated from the wind is that, as grim as the cost estimates and think tank forecasts can sometimes be, we are just getting started. "I think the good thing about lithium-air, lithium sulfur is that at least there are some options," Bruce said.

"There is somewhere we can go."

EPILOGUE

The inner-city neighborhood that surrounds General Motors's Detroit-Hamtramck Assembly Plant is a showpiece of postindustrial urban blight, the kind of place whose condemned architecture would work well on an ironic postcard delivering greetings from Murder City. But inside the plant's secure gates, one frigid morning the week before Christmas 2010, GM was initiating a new phase of rebirth. Under a fresh blanket of snow stood the first forty-five dealer-ready Chevrolet Volts. As five red transporter trucks rolled up, employees began clearing the windshields and rolling the cars on board, and the Volt's entry into society began. This first shipment was bound for greater New York City and Washington, D.C.; by the end of the week another three hundred or so Volts would also be dispatched to California and Texas. The Volt's public-relations team quickly posted photographs of the inaugural shipment online, and after clicking through the slide show, Chelsea Sexton wrote on Twitter what every veteran of the EV1 wars had to be thinking: "I'm liking this view of EVs on a transporter much better than the last time!"

The Volt rollout had really begun a couple of months earlier, with early media test drives of "salable" prototypes—the last cars to be built before official production, vehicles that to the untrained eye were indistinguishable from the ones that would go to customers. My drive came in October, and when I arrived at the Detroit airport on a Sunday afternoon, a GM engineer escorted me to a curbside row of Volts. As I settled into the driver's seat of a shiny black specimen, I was struck by how far the car had come since my spin in a hacked-up prototype a year and a half earlier. It shouldn't have been surprising that the Volt would end up as an attractive, fairly loaded production car, and yet in a way it was—probably because I had spent the previous three years thinking about the Volt as a science project rather than as a product that would eventually go on sale. As we drove away from the passenger pickup zone and reached a modest cruising speed, the Volt was frictionless and silent. Steering was silky and precise. The car launched from a stop with a punch that made it feel faster than it actually was. The next day, when I had a chance to get it to 85 mph on the freeway, it felt unshakable and nimble.

The car, in short, was fantastic. Yet every step of the Volt rollout process met with some form of backlash. The week of my test drive, the fury was fueled by an esoteric revelation about the nature of the car's blended power train: that in some circumstances, at high speeds with the battery depleted, the gasoline engine could connect (via a small electric motor) to the gearset that drives the wheels. I was in the room when GM engineers explained the arcane mechanics of this process to a group of about thirty reporters, and the news seemed pretty uncontroversial. But almost immediately a vastly oversimplified version of the story reached the outside world, and by lunchtime, the blogosphere had gone insane. A writer for Edmunds.com posted on Twitter: "Rumors are true. GM lied to the world. Volt's engine does power the car's wheels. It isn't a true EV as promised." A wave of geeky outrage rolled across the car blogs so quickly that soon *The New York Times* was addressing the matter.

GM's public-relations people should have been accustomed to this kind of reaction. When GM finally announced the Volt's $41,000 price tag in late July 2010, critics complained about the steep figure and declared the car, which was still nearly five months away from dealerships, a failure. An Op-Ed in *The New York Times* called the Volt "GM's Electric Lemon." Because of the Obama administration's bailout of GM, a revisionist history of the origins of the Volt soon took hold in certain political circles. According to this story, the Volt had been the Obama administration's idea, and the car's creation was a necessary condition of a government rescue of GM. In November, George Will wrote a column in which he called the Volt a "government brainstorm." The tax credits that support the Volt and other electrified vehicles were "bribes." President Obama was, in Will's sarcastic words, "Automotive Engineer in Chief," and the federal bailout of GM and Chrysler was a case of "government and its misnamed 'private sector' accomplices foist[ing] state capitalism on an appalled country." When *Motor Trend* announced that the Volt had won its Car of the Year award, Rush Limbaugh screamed that it would be the "end" of the magazine. "How in the world do they have any credibility?" he cried. "Not one has been sold."

In a response published online, *Motor Trend*'s Detroit editor, Todd Lassa, pointed out that no Volts had been sold because the car was not yet for sale, and told Limbaugh that if he would bother to drive the Volt, he might enjoy it. "Just remember: driving and Oxycontin don't mix," Lassa wrote.

Ultimately, though, the polarization over the Volt seemed largely confined to the spheres of auto journalism and professional political hackery. When GM auctioned for charity the second Volt to exit the assembly line, the winning bid of $225,000 came from a man with impeccable red-state credentials: Rick Hendrick, the North Carolina car dealer who owns the NASCAR Sprint Cup team Hendrick Motorsports. When production of the Volt began, on schedule, on November 11, the reception was mainly enthusiastic. Interest in the car had grown so intense that by early December GM had announced that it was looking for a way to "double or triple" Volt production. That same week, GM's new CEO, Dan Akerson—the former telecom executive who took the job when Ed Whitacre stepped down unexpectedly in August—announced that GM would soon add one thousand electric-vehicle engineering and development jobs on top of the approximately two thousand people already working worldwide on the Volt and its future brethren. Finally, the success of the "New GM's" initial public offering on Wednesday, November 17, signaled a new era for the company. Investors devoured 478 million common-stock shares, and GM raised $20.1 billion in the largest IPO in American history.

While the Nissan Leaf was a riskier proposition than the Volt, its arrival in late 2010 was drama-free. Although Nissan insisted that the Leaf and the Volt were completely different types of vehicles—one was a pure EV, the other a variation on the plug-in hybrid—the comparison of the two was inevitable, and the slow striptease that was the Leaf launch adhered to almost exactly the same schedule as the Volt's. The week after driving the Volt in Detroit, I joined a group of reporters at Nissan's North American headquarters outside Nashville, and on winding two-lane highways amid the horse farms of rural Tennessee, the Leaf was a pleasure to drive. On the highway, it easily sailed to the brink of 90 mph, and at such speeds it remained quiet, sturdy, and smooth. The reviews of the car were positive, and a month and a half later, Nissan gave a press conference in San Francisco's Civic Center Plaza to mark the world's first sale of a Leaf, to a Silicon Valley entrepreneur named Olivier Chalouhi.

That the Leaf was able to sneak into the marketplace with so little backlash suggests that electric cars aren't nearly as controversial as is General Motors—that the shrieking debate over the Volt had less to do with the underlying technology than it did with the fact that the Volt is a GM product. After all, by the end of 2010 the question wasn't whether the automobile would be electrified. It was how quickly that would happen, and what mark on the spectrum between pure gas power and pure battery power made the most sense. Other automakers may have been more cautious with electrification plans than either GM or Nissan, but they soon got on board. Ford, for example, announced that it would release a purely electric version of its Focus at the end of 2011, and that a plug-in hybrid would follow in 2012. Toyota announced that in 2012 it would begin selling a plug-in version of the Prius. At the Los Angeles Auto Show in November 2010, Mitsubishi revealed a North American version of its i MiEV electric car, which it said would arrive in the United States a year later. In September, Volkswagen AG chairman Martin Winterkorn had told *Der Spiegel* that excitement over hybrids would fade once people realized that it was a "bridge technology." "The next big step is the electric car," he said. Two months later, Audi declared its intent to lead the "premium" EV market by 2020. Even Porsche approved production of a hybrid supercar, the 918 Spyder, and announced that the car was only a first step in a large electrification program.

The auto industry's momentum was welcome news for companies such as A123 Systems, which in September 2010 opened a 291,000-square-foot automotive-battery plant in Livonia, Michigan. In a talking point surely designed to needle the competitors at EnerDel, the company claimed that the Livonia facility was "the largest lithium-ion facility in America." On opening day, the plant employed three hundred people, but A123 estimated that in about a year, between the Livonia plant and another Michigan facility, it would be responsible for creating some three thousand jobs in the state—jobs that, without the help of government incentives, would have been created overseas.

Because the company built the plant in part with their $249 million stimulus grant, the plant's opening ceremony became a political event. Michigan's governor, Jennifer Granholm, and Secretary of Energy Steven Chu joined Yet-Ming Chiang and the rest of A123's executives in Livonia that day, and President Obama called in to deliver a few remarks. "This is important," Obama told the gathered parties, "not just because of what you guys are doing at your plant, but all across America, because this is about the birth of an entire new industry in America—an industry that's going to be central to the next generation of cars." He made a point of contrasting his administration's nurturing stance toward companies like A123 with those that had come before. "For a long time, our economic policies have shortchanged cutting-edge projects like this one, and it put us behind the innovation race," he said.

In an interview with CNBC, A123's CEO, David Vieau, asserted his optimism about the company's prospects: in the 2012–2013 time frame, he said, the EV and plug-in market would begin taking off. By 2015, it would have expanded "dramatically, independent of the price of gas."

As 2011 began, the first mass-market electric cars since the dawn of the twentieth century were sitting in dealerships. The dream of founding a domestic lithium-ion battery industry has begun to yield real jobs and real factories. Still, the question remains: Will this last? Is this book the story of the beginning of something enduring—a new era for transportation, energy, and American high-tech manufacturing? Or will it turn out to be a portrait of another false start, an anomalous couple of years in which, once again, scientists and entrepreneurs and government attempted to seed an energy revolution, only to see those seeds die when "business as usual" resumed?

Judging by the major, bank-breaking commitments of automakers around the world—not to mention the eagerness of rising world powers such as China to dominate the new industry—the gradual electrification of the automobile appears inevitable. It won't happen immediately. Until battery technology improves still further, purely electric cars like the Nissan Leaf will remain limited in long-haul America. But as the Chevy Volt demonstrates, electrification isn't an all-or-nothing proposition—smart engineers will use the best battery technology available at any given time to displace as much petroleum as possible, and when those batteries run out, gasoline will continue to get the job done.

The most common argument one hears against cars like the Volt in the United States today is that gasoline still isn't expensive enough to justify the additional cost of a car that runs on a pricey lithium-ion battery. But that argument has a built-in expiration date. Battery cost has already begun to plummet: see, for example, the $32,780 sticker price of the Nissan Leaf, a car with a relatively large lithium-ion battery. Gas prices were relatively low at the time the Volt and the Leaf arrived in the world, but those prices will not last forever. The supply of oil is finite, and many well-informed observers believe that someday soon, it will be much too scarce a commodity to continue powering our lives in the way it does today.

What remains to be seen, of course, is who will profit most from this shift in the automotive industry. Young American companies like A123 and EnerDel face profoundly tough competition from giants like Panasonic—which is now working closely on automotive applications with Toyota and Tesla Motors, and which, according to Tesla's founder,

Martin Eberhard, is sitting on the next big battery breakthrough, the silicon anodes that as soon as 2013 will increase the energy density of lithium-ion batteries by as much as 30 percent.

The biggest threat to the hatchling American battery industry, however, could be politics. The day after the 2010 midterm elections, in which the Republicans swept the House of Representatives on an anti-Obama wave, could not have been pleasant for anyone with a stake in the emerging American advanced-battery industry. Soon after the election, Representative Fred Upton, a Michigan Republican, wrote to Steven Chu complaining about the portion of the stimulus funding awarded by the Department of Energy.

The industry advocate James Greenberger believes that this latest political shift will change the way the battery industry sells itself—expect to hear less about cutting carbon dioxide emissions and more about establishing energy security by developing alternatives to foreign oil—but that it shouldn't, in itself, deal the industry a deathblow. After all, the industry's greatest incubator, the stimulus funding, was a one-time deal. "The $2.4 billion of DOE grants to support advanced battery manufacturing and the electric drive supply chain was not funded by something called the Battery Package," Greenberger wrote on his blog. "It was funded by the Stimulus Package. That is an important distinction. Once the Stimulus Package funds were expended and the economy began to recover, there was never any realistic expectation that anything close to that level of funding for battery manufacturing was going to continue. Industry was going to have to become self-reliant and the business of advanced batteries self-sustaining."

But it is clear that the brief window in which any kind of comprehensive climate-change legislation was possible—any kind of wide-ranging effort to penalize carbon emissions and therefore boost clean alternatives, including electric cars—has been closed, and it will remain closed for years to come. The EPA is still charged with regulating greenhouse gases, but Republicans in the House of Representatives will try to stop the agency from doing so. The best tool the government has for urging on advances in clean-car technology is probably Corporate Average Fuel Economy (CAFE) standards, and in fact, in October 2010, the Environmental Protection Agency and the National Highway and Transportation Safety Administration signaled that CAFE standards could rise from the 35.5 mpg average set for 2016 to as high as 62 mpg by 2025. Still, for the next two years, Washington will probably become a more hostile place for clean-energy interests.

In a speech at the National Press Club in late November 2010, Steven Chu went on the offense against any possible anti-energy-research agenda coming with the next Congress. He explained that the budget for energy research has steadily declined since the 1970s—that today, only 0.14 percent of the federal budget is allocated for energy research and development. The stimulus should be a "down payment" on a long-term program of energy R&D. "The question is, post-stimulus, are we going to return to this downward trend or are we going to do something about it?"

A presentation Chu distributed to the audience explained that some level of government direction of the private sector is necessary because the benefits of clean-energy technology—clean air, better national security, less risk of dangerous climate change, stability of energy prices—are "neither recognized nor rewarded by the free market." High-risk, high-reward energy research of the kind that could deliver major breakthroughs is too risky for most private corporations. "And quite frankly," Chu said, "a lot of the new technologies could displace an embedded way of doing business, and could be met with resistance; therefore the government has to say, 'This is the path we should be going in for our long-term future prosperity.'"

Chu's speech was titled "The Energy Race: Our New Sputnik Moment." He admitted that the Sputnik analogy was trite, but said that in this case it should perhaps be taken seriously. The reason: the United States is demonstrably on the verge of losing its position as the world leader in science and technology. The nation most eager to claim that leadership is, of course, China. "The U.S. has been for well over a century the greatest innovation machine in the world," he said. "While it did not invent the automobile, it took the invention and processed it into something that was not seen in the world before . . . I say today this leadership is at risk."

But Chu admits that the analogy with the Sputnik era extends only so far. The world needs "a new industrial revolution," he believes. If another country leads that industrial revolution, then, all things being equal, the result will still be good for the planet—Americans will just be buying solar panels and carbon-capture technology and advanced batteries from overseas. And if the budding American energy-storage industry fails, it's not going to be an existential threat to the United States. It would, however, be a tremendous lost opportunity, a failure to participate in what promises to be one of the greatest industries of the coming century.

APPENDIX

Global Lithium Reserves and Identified Resources

Based on the latest U.S. Geological Survey estimates as of January 2011. Reserves are mineral sources that can today be economically and legally extracted; identified resources are known mineral deposits. All figures are in metric tons.

RESERVES		IDENTIFIED RESOURCES	
Chile	7,500,000	Bolivia	9,000,000
China	3,500,000	Chile	7,500,000
Argentina	850,000	China	5,400,000
Australia	580,000	United States	4,000,000
Brazil	64,000	Argentina	2,600,000
United States	38,000	Brazil	1,000,000
Zimbabwe	23,000	Congo	1,000,000
Portugal	10,000	Serbia	1,000,000
Total	12,565,000	Australia	630,000
		Canada	360,000
		Total	32,490,000

NOTES

I: The Electricians

9–10 Thales of Miletus . . . Benjamin Franklin: Jonnes, *Empires of Light*, pp. 17–49.

10 The battery was the accidental fruit: The main source for Volta's story is Pancaldi, *Volta*, pp. 178–207.

11 paper by the English chemist William Nicholson: "It has appeared to me . . . that a machine might be constructed also capable of giving numberless shocks at pleasure, and of retaining its power for months, years, or to an extent of time of which the limits can be determined only by experiment." Ibid., p. 199.

11 "electricity excited by the mere mutual contact": Jonnes, *Empires of Light*, p. 32.

12 "loud detonations": Pancaldi, *Volta*, p. 215.

12 Volta called his invention: In 1803, Humphry Davy pretty much settled the matter when he used the term "galvanic battery" in a paper. Today the English still call it the pile, but to most of the rest of us, the electrochemical cell has been called the "battery" ever since.

12 "the last great discovery": Quoted in Pancaldi, *Volta*, p. 211.

12 "magnificent instrument of philosophic research": Quoted ibid., p. 273.

12 called Volta "immortal": Quoted ibid., p. 259.

12 "opened to man a new and incomparable source of energy": Quoted ibid., p. 273.

12 Volta earned such effusive praise: In Italy, Volta became a national hero. On the hundredth anniversary of the battery, an Italian trade group hired Giacomo Puccini to write commemorative music, and the result was a piano piece Puccini called "The Electric Shock." Then in 1927, on the centennial of Volta's death, Italy's

Fascist government threw a massive celebration in Como, Volta's birthplace. Mussolini was "honorary president" of the proceedings. From fourteen countries, sixty-one physicists gathered in Como, among them the giants: Niels Bohr, Max Planck, Ernest Rutherford, Werner Heisenberg, Enrico Fermi. On postage stamps issued for the centennial, "Alessandro Volta was portrayed in the pose of an ancient Roman. Volta's electric battery was made to look like the bundle of elm branches containing an axe that was the symbol of the fascist regime" (ibid., p. 264).

13 Hans Christian Oersted: Pancaldi, *Volta*, pp. 233–34.

13 As the battery steadily improved: Throughout the nineteenth century, successive generations of more powerful primary, or nonrechargeable, batteries arrived. In 1836, John Frederic Daniell, an English chemist, invented the first major improvement on Volta's pile, a cell that used a zinc electrode and a copper electrode, each dipped in separate containers that were filled with sulfate solutions, and then connected to one another with a salt bridge. In 1844, another Englishman, William Robert Grove, concocted a cell that used zinc and platinum electrodes to reach 1.9 volts. In 1866, Georges Leclanché delivered a zinc-carbon primary (nonrechargeable) battery whose design would in time lead to the first "dry" cell, which used a paste rather than the traditional liquid solution for an electrolyte. See Schallenberg, *Bottled Energy*, and Schlesinger, *The Battery*, for a more detailed account.

14 And so in 1898, he began studying: The section on Edison's struggle with the battery draws on three main sources: Josephson, *Edison*; Schallenberg, *Bottled Energy*; and Schiffer et al., *Taking Charge*. As Josephson wrote, "From 1900 on, he had eyes for nothing but the 'miniature reservoir of electric force' that he must create" (*Edison*, p. 407).

14 "I don't think Nature": Josephson, *Edison*, p. 407.

15 "The number of experiments": Ibid., p. 409.

15 In reality, he was not working blindly: Schallenberg, *Bottled Energy*, pp. 353ff.

15 "the final perfection of the storage battery": Thomas A. Edison, "The Storage Battery and the Motor Car," *North American Review* 175 (1902): 1–4.

15 "a featherweight and inexhaustible": Ritchie E. Betts, "Faster than the Locomotive," *Outing: An Illustrated Magazine of Sport, Travel, Adventure & Country Life* 339 (1901–1902).

16 "revolutionized the world of power": Josephson, *Edison*, p. 415.

16 "At last the battery is finished": Quoted ibid., p. 421.

17 Today we know: Armstrong, R. A., G.W.D. Briggs, and M. A. Moore. "The Effect of Lithium in Preventing Iron Poisoning in the Nickel Hydroxide Electrode," *Electrochimica Acta* 31, no. 1 (1986): 25–27.

17 In 1800, a Brazilian chemist: José Bonifácio de Andrada e Silva documented his discovery in *Allgemeines Journal der Chemie* in 1880. See Mindat.org's page on Utö for detail on the site: www.mindat.org/loc_3194.html.

17 Johan August Arfwedson: Encyclopædia Britannica Online: "lithium (Li)," www.britannica.com/EBchecked/topic/343644/lithium.

17–18 Lithium therapy became popular . . . "It takes the ouch out of grouch": El-Mallakh and Jefferson, "Prethymoleptic Use of Lithium;" El-Mallakh and Roberts, "Lithiated Lemon-Lime Sodas."

18 began giving heart-disease patients lithium chloride: El-Mallakh and Jefferson, "Prethymoleptic Use of Lithium," p. 129.

18 the Australian psychiatrist John Cade: Cade, "Lithium Salts in the Treatment of Psychotic Excitement," pp. 349–52.

18 lithium affects neurotransmitters: B. Corbella and E. Vieta, "Molecular Targets of Lithium Action," *Acta Neuropsychiatrica* 15 (2003): 316–40.

18 stimulate brain-cell growth: Moore et al. "Lithium-Induced Increase in Human Brain Grey Matter."

18 compared suicide rates and lithium levels: Ohgami et al., "Lithium Levels in Drinking Water and Risk of Suicide."

19 a Canadian psychiatrist suggested: Young, "Invited Commentary."

21 Even when it is generated by: Numerous studies have compared "well-to-wheel" emissions for electric cars, plug-in hybrids, and internal combustion engines. For an overview, see Sherry Boschert, "Well-to-Wheels Emissions Data for Plug-In Hybrids and Electric Vehicles: An Overview," www.sherryboschert.com/Downloads/Emissions%5B9%5D.pdf.

2: False Start

22 The clouds of smog: Details on the smog crisis are drawn from "Menace in the Skies," *Time*, January 27, 1967. For a comprehensive history of LA's smog problems, see Jacobs and Kelly, *Smogtown*.

23 backlash against the internal combustion engine: Doyle, *Taken for a Ride*, pp. 55ff.

23 outright banning of the internal combustion engine: Ibid., p. 55. California state senator Nicholas C. Petris proposed one such bill in 1969.

23 By then, the geopolitics: In researching the oil situation in the 1960s and 1970s, I've drawn on Daniel Yergin's *The Prize*, the definitive history of oil, which delivers a comprehensive account of those tumultuous decades. For a concise history of oil and an immediate, journalistic account of the first oil crisis, see also Sampson, *The Seven Sisters*.

24 "It was a decisive change": Yergin, *The Prize*, p. 573.

25 In 1967, Neil Weber and Joseph T. Kummer: Weber and Kummer, *Proceedings of the Annual Power Sources Conference* 21 (1967): 37.

26 intercalation compounds: Huggins, *Advanced Batteries*, p. 61.

27 A yellowed black-and-white photo: van Gool, *Fast Ion Transport in Solids*.

29 tantalum disulfide: Interview with Michael Stanley Whittingham, SUNY Binghamton, October 30, 2000, http://authors.library.caltech.edu/5456/1/hrst.mit.edu/hrs/materials/public/Whittingham_interview.htm.

29 an excellent conductor of electricity: This was a major advantage. Almost all other electrode materials have to be mixed with a healthy dose of carbon black to

make them conduct electricity; when as much as 20 percent of the electrode is taken up by carbon, that's 20 percent less active electrode material you can fit in, 20 percent less real estate for lithium ions—the real charge-carrying workhorses of the battery. Titanium disulfide was such a good conductor that they could skip the carbon black entirely.

31 landmark paper on the $LiTiS_2$ battery: Whittingham, "Electrical Energy Storage and Intercalation Chemistry."

32 "one of today's hottest items": "The Best Growth Business," *Forbes*, May 15, 1975.

32 "despite present—and formidable—problems": "Car of the Future," *Forbes*, October 15, 1976.

32 "After a hiatus of almost 50 years": "New Batteries Are in the Running," *Chemical Week*, December 1, 1976.

32 "Given two major trends": James Flanigan, "Does Exxon Have a Future?" *Forbes*, August 15, 1977.

33 a Vienna hotel room: Yergin, *The Prize*, p. 583.

34 In a presentation: "Down to Earth Talk on Far Out Ideas," *Chemical Week*, February 22, 1978, p. 42.

34 the company spent $1.2 billion: Richard I. Kirkland, Jr., and Susan Kuhn, "Exxon Rededicates Itself to Oil," *Fortune*, July 23, 1984.

35 "What Exxon is saying": "Exxon Puts a Tiger in Your Electric Motor," *Economist*, May 26, 1979, p. 111.

35 "We're not finding as much oil": "Interview with Clifton Garvin: 'The Quicker We Get at Synthetic Fuels, the Better We're Going to Be,'" *BusinessWeek*, July 16, 1979, p. 80.

35 "That may be just as well": "Here Come the Electrics," *Fortune*, October 22, 1979, p. 24.

36 Nothing, that is, except oil: Hamlen believes that the death knell for the Battery Division was sounded when a potentially huge deal fell through. The group was finding steadily larger applications for their batteries, in particular a solar rechargeable desk clock that caught the eye of Charles Tandy, founder of the eponymous, now-forgotten computer company. Tandy had just bought RadioShack; Hamlen had good reason to believe he wanted to buy fifty thousand lithium-battery-powered clocks. Then, on November 29, 1978, Tandy died at age sixty of a heart attack. The sale was finished.

Exxon Enterprises became a case study in how corporate diversification can go wrong. As Hollister B. Sykes, formerly of Exxon Enterprises, wrote in the *Harvard Business Review* in May 1986, "The high proportion of R&D ventures in our portfolio greatly increased our risk of failure and stretched out the time from start-up to projected sales. Because most corporations go through cycles in their base businesses, unprofitable operations not in the mainstream are especially vulnerable. Exxon was no exception. The steep slump in the consumption of oil products and natural gas from 1979 to 1982 caused concern. Along with the cutback in Exxon's base business operations, we either sold or liquidated most of our smaller ventures."

37 "We're not interested": Kirkland and Kuhn, "Exxon Rededicates Itself to Oil."

38 "Without a market": "The Slow, Sure Advance Toward Better Batteries," *Chemical Week*, November 28, 1984.

3: The Wireless Revolution

39 Goodenough is now a professor: Many of Goodenough's biographical details come from his book *Witness to Grace*.

41 Douglas H. Ring and William Roe Young: Private Line, "Cellular Telephone Basics," January 1, 2006, www.privateline.com/mt_cellbasics/ii_cellular_history/.

42 On April 3, 1973: Stephanie N. Mehta, "Cellular Evolution," *Fortune*, August 23, 2004, p. 80.

42 Process of elimination: Bernadette Bensaude-Vincent and Arne Hessenbruch, "Interview of John B. Goodenough," March 2001, http://authors.library.caltech .edu/5456/1/hrst.mit.edu/hrs/materials/public/Goodenough/Goodenough _interview.htm.

43 before it started to crumble: During discharge, lithium ions swim from the anode to the cathode, where they embed themselves into the crystalline lattice of the cathode. When the battery is being charged, those lithium ions flee the cathode and head back to the anode. The whole project rested on whether the microscopic Jenga puzzle that is a crystalline lattice could stay erect even when half of the lithium ions docked within it were sucked away.

44 An excellent 4 volts: K. Mizushima, et al., "Li_xCoO_2 ($0<x<-1$)."

44 But that was just in the United States: This paragraph draws on Tom Farley's article "Mobile Telephone History," *Telektronikk* 3/4 (2005): 22–34.

45 From a car outside Soldier Field: "Cellular Mobile Phone Debut," *New York Times*, October 14, 1983.

45 approximately $3,000 for an Ameritech car phone: Ibid.

45 first handheld mobile phone: Ted Oehmke, "Cell Phones Ruin the Opera? Meet the Culprit," *New York Times*, January 6, 2000.

46 "an adjunct of the microelectronics revolution": Jonathan Greenberg, "The Battery Boom," *Forbes*, November 8, 1982.

46 "Mercury, a toxic metal": Andrew Pollack, "Battery Pollution Worries Japanese," *New York Times*, June 25, 1984.

46 "swallowing issue": In the summer of 2010, the "swallowing issue" reemerged, and the culprit was button-cell primary lithium batteries. See Tara Parker-Pope, "For Very Young, Peril Lurks in Lithium Cell Batteries," *New York Times*, May 31, 2010.

48 "They kind of got left in the dust": Some in the industry suspected that simple greed guided the American battery companies' lack of interest in rechargeables. After all, primary battery manufacturers made their living selling products you

use and throw away. But there's more to it than that. A 2005 study commissioned by a program within the National Institute of Standards and Technology identifies an array of reasons lithium-ion manufacture didn't take hold in the United States. Among them: The Japanese companies who led the rise of the advanced rechargeable battery weren't just battery manufacturers—they were *electronics* manufacturers, for whom having an in-house battery manufacturer that could engineer specific power sources for particular devices, leading to overall better and more profitable gadgets, made perfect sense. And because the gadgets that were driving demand for advanced batteries were being designed in Japan, by companies that had their own battery divisions, it would have been a challenge for the American manufacturers to get into this market even if they had dearly wanted to. Add to all this the cultural differences between American and Japanese business—most significant, the focus on short-term gains versus long-term success—and it's not surprising that Japan ran away with the lithium-ion battery market. Ralph J. Brodd, "Factors Affecting U.S. Production Decisions: Why Are There No Volume Lithium-Ion Battery Manufacturers in the United States?" June 2005, www.atp.nist.gov/eao/wp05-01/contents.htm.

48 According to Sony's official history: www.sony.net/SonyInfo/CorporateInfo/History/SonyHistory/index.html.

48 a hostile takeover bid: States News Service, April 16, 1986.

50 In February 1990: "Sony Subsidiary to Begin Shipping Lithium Ion Battery," Japan Economic Newswire, February 14, 1990.

51 The founder was Rudi Haering: Interviews with Jeff Dahn; Ann Shortell Vansun, "Recharging Moli's Battery," *Vancouver Sun*, September 26, 1991.

52 By the spring of 1986, Moli: David Climenhaga, "Firm Finds Battery Has Drawing Power," *Globe and Mail*, March 13, 1986.

52 held a public stock offering: "Moli Starts Producing at Vancouver Plant," *Battery and EV Technology* 19 (1994).

52 in 1988, Moli's first battery: "Moli Energy Introduces New High-Energy, High-Voltage Rechargeable," *PR Newswire*, December 15, 1988.

52 an NTT phone equipped with a Moli battery: Ann Shortell Vansun, "The Question of Quality," *Vancouver Sun*, October 1, 1991.

53 Moli laid off 56 of its 192 employees: "Moli Energy Lays off 56," *Financial Post* (Toronto), October 2, 1989.

53 Mitsui stepped in: Anne Fletcher, "Mitsui & Co. to Take Control of Troubled Moli," *Financial Post* (Toronto), March 12, 1990.

53 Motorola sold $180 million worth: The figures in this paragraph come from "The Role of NSF's Support of Engineering in Enabling Technological Innovation; Phase II," Chapter 4, SRI International, May 1998, www.sri.com/policy/csted/reports/scrdt/techinZ/chp4.html.

54 "The competition is becoming fierce": David Thurber, "Dull but Durable, Rechargeable Batteries Enjoy Surging Demand," Associated Press, March 10, 1992.

54 "While electronics manufacturers": "The Battery Market; Recharged," *Economist*, May 2, 1992.

54 Beginning in 1992, Sony offered: "Sony's 'Handycam Station,'" *Consumer Electronics*, September 7, 1992.

55 In 1993, there were thirteen million cell phones: Peter Haynes, "The End of the Line," *Economist*, October 23, 1993.

55 by some 25 percent a year: Ibid.

55 electromagnetic fields came in at number one: David Kirkpatrick, "Do Cellular Phones Cause Cancer?" *Fortune*, March 8, 1993.

55 a guest on *Larry King Live*: "The Wireless Wonder," *Forbes*, April 26, 1993.

55 In 1994, *Electronic Engineering Times*: Martin G. Rosansky and Ian D. Irving, "Lithium-Ion Offers Better Choices," *Electronic Engineering Times*, May 23, 1994.

55 In Japan, manufacturers were scrambling: Hisayuki Mitsusada, "Lithium Ion Set to Recharge Firm's Sales," *Nikkei Weekly*, August 21, 1995.

56 total Japanese dominance: Even as its competitors got into the lithium-ion business, Sony was so far ahead that on November 4, 1995, when a fire in Sony-Energytec's Koriyama factory caused a train wreck in the global battery supply chain, none of Sony's competitors had anywhere near the capacity to take up the slack. Sony had been making three million cells a month in Koriyama; its Tochigi factory still wasn't running when the fire erupted. As a Japanese industry analyst noted at the time, global demand for lithium ion that year would be forty million, and even with A&T, Matsushita, and Sanyo running full bore, the Sony fire meant that the market would probably be ten million cells short. Mark LaPedus, "Fire Chokes Battery Supply—Sony Halts Output at Lithium-Ion Plant," *Electronic Buyers' News*, November 13, 1995.

57 Weighing only 3.1 ounces: Scott Hume, "Motorola's Big Push for Smallest Phone," *Adweek*, January 8, 1996.

57 "will WEAR rather than carry it": "Motorola Unveils European StarTAC Cellular Telephone," Motorola press release, March 28, 1996.

57 "A big cellular phone": Paul Hochman, "Mine Is Smaller Than Yours," *Forbes*, November 18, 1996, p. 136.

57 "twenty indispensable luxuries": Esther Wachs Book, "'Tis the Season to Covet," *Fortune*, December 23, 1996, p. 258.

57 a choice gangsta fashion item: Dana Kennedy, "Deadly Business," *Entertainment Weekly*, December 6, 1996, p. 34.

57 sixth greatest gadget: Dan Tynan, "The 50 Greatest Gadgets of the Past 50 Years," *PCWorld*, December 24, 2005.

58 more than 120 million: Richard Ernsberger, Jr., and Judith Warner, "War for the World," *Newsweek*, February 10, 1997.

58 two each second: Manuel Del Castillo and Henry Valenzuela, "The Role of Microwave Technologies in the Wireless Revolution," *Microwave Journal*, September 1, 1998.

58 200 to 300 percent per year: "Battery Manufacturers Charge Ahead, Despite Impending Glut," *Nikkei Weekly*, July 28, 1997.

58 Internet-capable phone called i-mode: "The World in Your Pocket," *Economist*, October 9, 1999.

58 The merging of the cell phone: "The Conquest of Location," ibid.

58 A piece in *Time*: Adam Cohen, "Wireless Summer," *Time*, May 29, 2000.

58 "Perhaps the more worrisome outgrowth": Stephen Baker, Neil Gross, Irene M. Kunii, and Roger O. Crockett, "The Wireless Internet," *BusinessWeek*, May 22, 2000.

59 "foot-soldiers of the digital revolution": "Hooked on Lithium," *Economist*, June 22, 2002.

4: Reviving the Electric Car

60 sold NuvoMedia for $187 million: "Valley Techs Tackle Electric Car," *Australian*, July 11, 2006.

62 They soon arrived at the design for their battery pack: When I told Eberhard that powering a car with 6,831 laptop cells wired together sounded incredibly make-shift, he replied, "When you say it like that, that's true. That's like saying a car is a whole bunch of sheet metal and bolts all stuck together."

64 Even the editorial board: "Go Speed Racer!" *New York Times*, July 23, 2006.

65 one of the best inventions of 2006: "Best Inventions 2006: Batteries Included," *Time*, November 13, 2006.

65 some $6 billion a year on benefits: "GM Will Reduce Hourly Workers in U.S. By 25,000," Danny Hakim, *New York Times*, June 8, 2005.

65 the company lost $1.1 billion: Sharon Silke Carty and James R. Healey, "GM Takes $1.1B Hit in First Quarter," *USA Today*, April 20, 2005.

67 earning a place: Halberstam, *The Reckoning*, pp. 741–42.

68 a long time coming: For a deep account of GM's downfall, see Paul Ingrassia's *Crash Course*.

68 tellingly called iCar: Holstein, *Why GM Matters*, p. 131.

70 possible career killers: A GM engineer named Kenneth Baker confronted a similar situation in 1990 when he was asked to lead the EV1 program. He had led the doomed Electrovette program in the late 1970s, and over the years he came to feel that its failure had stalled his career. When he was recruited to do the EV1, he was afraid that it could be his undoing. See Michael Shnayerson's excellent *The Car That Could* for a detailed account of the EV1 saga.

72 The language Wagoner used: A GM executive would argue that the hydrogen fuel-cell research that the company conducted in the years after the demise of the EV1 shows that they never did turn away from electrically driven vehicles.

74 ten times the attention: Tom Walsh, "GM's Message to the World—Don't Count Us Out," *Detroit Free Press*, January 10, 2007.

74 "not a public relations ploy": Lindsay Brooke, "All the Technology Needed for 100 M.P.G. (Batteries Not Included)," *New York Times*, January 7, 2007.

5: The Blank Spot at the Heart of the Car

75 The idea that electricity could be used as a locomotive force: This section on the early history of the electric car owes a considerable debt to Schiffer's thorough and eminently readable *Taking Charge*. For a more academic take on the same period, see Kirsch's *The Electric Vehicle and the Burden of History*, which also proved valuable in assembling the historical passages of this book.

76 "Men fell on their knees": Schiffer, *Taking Charge*, p. 51.

76 "The day does not seem so very far distant": Ibid., p. 35.

78 "a hospital full of sick dogs": Quoted ibid., p. 64.

79 "That any firms lasted past the war": Ibid., p. 159.

79 Not until air pollution became an issue: The document that establishes this link most clearly is the report of the Department of Commerce's 1967 panel on the problem: *The Automobile and Air Pollution: A Program for Progress*. Details about the experimental electric city cars of that same era (the GM XP512E, the AMC Amitron, and the like) are drawn from Shacket, *The Complete Book of Electric Vehicles*.

80 The Impact concept car had grown out of the Sunraycer: The definitive source on the CARB zero-emissions mandate and the saga of the EV1, which I rely on in my capsule history of this period, is Shnayerson, *The Car That Could*.

81 "General Motors is preparing": Matthew L. Wald, "Expecting a Fizzle, G.M. Puts Electric Car to Test," *New York Times*, January 28, 1994.

82 lobbying the state of California: The story of the electric-vehicle wars of the 1990s is sprawling, complex, fascinating, and told very well in a number of books. Shnayerson provides the view from inside General Motors, which in those years had passionate engineers making an urgent good-faith effort to develop the EV1 while at the same time executives and lobbyists were fighting California's zero-emission mandate and effectively working against the success of the electric car. See Doyle's *Taken for a Ride* for a full account of the conspiracy charges of the era, in which the Sierra Club Legal Defense Fund sued the Big Three, accusing them of colluding to "hinder the introduction of electric vehicles" (pp. 305–23). The same chapter includes evidence that "the Big Three used USABC [United States Advanced Battery Consortium] to limit or suppress battery technology development." Boschert's *Plug-in Hybrids* explains the theory that Chevron may have intentionally limited automakers' access to nickel-metal-hydride batteries, which when installed in the final generation of EV1 extended the driving range to 160 miles. On the subject of nickel-metal-hydride batteries, see Shnayerson for a portrait of Stan Ovshinsky, the legendary inventor behind the nickel-metal-hydride battery, which many argue should have made the EV1 a truly attractive vehicle. *Who Killed the Electric Car?* is, of course, an essential source for understanding the story of the EV1.

82 "a free RAV4-EV plus a check": James R. Healey, "California May Soften Electric Car Mandate," *USA Today*, June 2, 2000.

82 the California Electric Transportation Coalition commissioned a study: "The Current and Future Market for Electric Vehicles," Green Car Institute, 2000.

82 Starting in 2002, Toyota: "Auto Emission Rules in California are Forcing Changes," Danny Hakim, *New York Times*, July 22, 2002.

83 Bereisa estimates that GM lost $1 billion: The claim that GM lost $1 billion on the EV1 is contentious. "At the time they stopped production entirely, the total cost was being reported as some $700 million," Chelsea Sexton wrote me in an e-mail. "How do you spend another $300 million on a car program that's no longer in production?" Sexton argues that creative accounting is at work here. "Whatever the number is, all sorts of other things would be assigned to it—the Chevy S-10 electric program, which ran parallel to EV1, all the lobbying costs to fight the mandate, and the full costs of the various technologies that are now being deployed in many other programs today. It's a common industry practice to bury all that in one program that's deemed a loss, so the other programs aren't burdened with those development costs." From GM's perspective, negative $1 billion is, in this case, an attractive number. "It makes a tidy little story about why the program was cancelled," Sexton wrote.

85 laptops powered by Sony lithium-ion batteries began catching fire: The details of the first incidents with Sony batteries and the Sony recall are drawn from Damon Darlin, "Dell Will Recall Batteries in PC's," *New York Times*, August 15, 2006.

85 on the blog Engadget: Cyrus Farivar, "Fire-Retardant Sleeves for Your Laptop," Engadget.com, September 16, 2006.

89 "It's like being in a conventional car at seventy miles an hour": Video for Greenfuelsforecast.com, www.youtube.com/watch?v=A17JrjXYcxs.

90 Tesla had had a rough couple of years: Michael V. Copeland, "Tesla's Wild Ride," *Fortune*, July 2008.

91 a "death watch" going: *EV World*, August 18, 2008.

91 "As a company, we do not have an official death watch": Jennifer Kho, "Toyota's Reinert Talks 'Death Watch' on Three Electric Cars," Greentechmedia.com, August 28, 2008.

91 Three weeks earlier: Holman Jenkins, Jr., "What Is GM Thinking?" *Wall Street Journal*, July 2, 2008.

6: The Lithium Wars

95 Goodenough and Padhi decided to present their results: Padhi, Nanjundaswamy, and Goodenough. "LiFePO$_4$."

96 "The phosphate was perfect": Armand continued, "It was poised at 3.5 volts. The electrolyte could be stable. The phosphate groups make the structure very sturdy, so it's not going to lose oxygen like the cobalt oxides and so on. And iron and

phosphate are very economical materials. It has everything. The only disadvantage is that it needs to be in very small particles. Small particles are hard to pack densely into a cathode, but this is probably the price to pay for safety."

97 In October 2002, Chiang's group: Chung, Bloking, and Chiang, "Electronically Conductive Phospho-olivines as Lithium Storage Electrodes."

97 Goodenough's old collaborator Michael Thackeray: Thackeray, "Lithium-ion Batteries: An Unexpected Conductor."

98 Authored by Armand and two colleagues: Ravet, Abouimrane, and Armand, "From Our Readers: On the Electronic Conductivity of Phospho-olivines as Lithium Storage Electrodes."

99 Armand's response was fundamentally flawed: Chiang's reply in *Nature Materials* continued, "Finally, although Ravet *et al.* speculate that our results are due to a coincidence of artefacts"—the carbon coating and the presence of Fe_2P—"they curiously do not observe these same artefacts in experiments that purport to reproduce ours. Despite this inconsistency in the argument, they contend that their results disprove ours. It is more likely that there are simple differences in experimental procedures."

100 Nazar's paper was published in 2004: Herle et al., "Nano-network Electronic Conduction in Iron and Nickel Olivine Phosphates."

100 "We also published a paper around that same time": Chung and Chiang, "Microscale Measurements of the Electrical Conductivity of Doped $LiFePO_4$."

100 the press was eager to listen: "New Type of Battery Offers Voltage Aplenty—at a Premium," *Wall Street Journal*, November 2, 2005; Jennifer Kho, "Battery Pumps Up Power Tools," *Red Herring*, November 1, 2005; Efrain Viscarolasaga, "A123 Charged and Ready to Hit Target Markets," *Mass High Tech*, November 14, 2005.

100 "We think this is equivalent": Viscarolasaga, "A123 Charged and Ready."

101 Hydro-Québec sent A123 a warning: Defendant's Sur-Reply in Opposition to Plaintiff's Motion to Reopen Case, *A123 Systems, Inc., v. Hydro-Québec*, No. 1:06-CV-10612-JLT, August 28, 2009, U.S. District Court, District of Massachusetts.

101 On April 7, 2006, the company filed an action: Complaint and Jury Demand, *A123 Systems, Inc., v. Hydro-Québec*, No. 06CV10612, April 7, 2006, U.S. District Court, District of Massachusetts.

101 On September 8, they requested a reexamination: Defendant's Opposition to Plaintiff's Motion to Reopen Case. *A123 Systems, Inc., v. Hydro-Québec*, No. 1:06-CV-10612-JLT, July 24, 2009, U.S. District Court, District of Massachusetts.

101 The University of Texas stepped in: Robert Elder, "Legal Fight over UT Patents Stretches On," *Austin American-Statesman*, October 15, 2006.

102 "a catch-penny, a sensation": *The Electrician*, February 17, 1883.

103 did some belligerent gloating: Jim Henderson, "Professor Is Mired in Patent Lawsuit; Visitor Accused of Lifting Research," *Houston Chronicle*, June 5, 2004.

103 a note that Okada had faxed: Ibid.

103 "He is a liar": Ibid.

104 In 2006, Nazar published another paper: Ellis et al., "Nanostructured Materials for Lithium-ion Batteries."

104 Then in 2008: Makimura et al., "Layered Lithium Vanadium Fluorophosphate."

104 Chiang hit back: Meethong et al., "Aliovalent Substitutions in Olivine Lithium Iron Phosphate and Impact on Structure and Properties."

104 Nazar and her colleagues: Ellis et al., "Comment on 'Aliovalent Substitutions in Olivine Lithium Iron Phosphate and Impact on Structure and Properties.'"

105 By January 2007, the U.S. Patent and Trademark Office: A123 Systems, June 30, 2010, Form 10-G (filed August 11, 2010), http://ir.a123systems.com/financials.cfm.

105 free to move forward: In June, A123 filed a motion to reopen their case, but in September the judge denied it. A123 is appealing the decision; they want the court to declare that their products do not infringe on the University of Texas's amended and narrower patents. And in May 2010, Hydro-Québec and the University of Texas filed their second amended complaint to reflect the changes in the patents. (China BAK Battery has been dropped from the suit, and A123 has indemnified Black & Decker.) At the time of this writing, a pretrial hearing had been scheduled; A123 wrote in an SEC filing that, if either lawsuit were to go forward, they would expect it to take two or more years to go to trial. The SEC filing also said this: "Regardless of the ultimate outcome of the litigation, it could result in significant legal expenses and diversion of time by our technical and managerial personnel. The results of these proceedings are uncertain, and there can be no assurance that they will not have a material adverse effect on our business, operating results, and financial condition."

106 In October 2008, Nippon Telegraph and Telephone settled: "NTT Settles Suit with U.S. University, Canadian Firm," Jiji Press Ticker Service, October 6, 2008.

106 Hydro-Québec transferred to its venture capital arm: William Marsden, "Hydro-Québec's Battery Goes Dead," *Gazette* (Montreal), September 3, 2005.

106 Then in 2001, Hydro-Québec signed over 50 percent: Ibid.

107 newly formed Montreal-based company: Alison MacGregor, "New Battery Firm Charges Ahead," *Gazette* (Montreal), October 16, 2001.

107 In September 2002, Avestor: Marsden, "Hydro-Québec's Battery Goes Dead."

107 AT&T installed seventeen thousand: Tyler Hamilton, "The Ugly Side of Next-Gen Energy Storage," *Clean Break*, January 16, 2008.

107 AT&T cable boxes equipped with Avestor batteries: Rick Barrett, "AT&T Pulling Batteries; Device Blamed in U-verse Equipment Cabinet Blast in Tosa," *Milwaukee Journal Sentinel*, January 17, 2008.

107 "Normally, we would work with a vendor": Linda Haugsted, "AT&T Will Replace Batteries After Fires," *Multichannel News*, January 21, 2008.

7: The Brink

110 Toyota became the biggest automaker: "Toyota Surpasses GM in Sales Amid Global Automotive Slump," *Toronto Star*, January 22, 2009.

110 the company lost $30.9 billion: Greg Keenan, "Losses Force GM to Question Its Future," *Globe and Mail* (Canada), February 27, 2009.

110 this time in a Volt prototype: Andrew Neather and Boris Johnson, "In Search of Electric Cars," *Evening Standard* (London), December 5, 2008.

111 GM's one hundredth anniversary party: Jim Motavalli, "G.M. Tones Down the Volt," *New York Times*, September 21, 2008.

113 "The most talked-about announcement": Kendra Marr, "GM Puts a Charge in Auto Show," *Washington Post*, January 13, 2009.

113 "Our timing had been set": Steven Rattner, "The Auto Bailout: How We Did It," *Fortune*, October 2009. The article was later expanded into the book *Overhaul: An Insider's Account of the Obama Administration's Emergency Rescue of the Auto Industry* (Boston: Houghton Mifflin Harcourt, 2010).

115 And on March 19, 2009: "Remarks at Southern California Edison's Electric Vehicle Technical Center in Pomona, California," www.gpoaccess.gov/presdocs/2009/ DCPD200900170.pdf.

117 Bob Lutz sent reassurances: March 31, 2009, http://gm-volt.com/2009/03/31/bob -lutz-volt-will-survive-and-prosper/.

117 strict new nationwide fuel-economy standards: John M. Broder, "Obama to Toughen Rules on Emissions and Mileage," *New York Times*, May 18, 2009.

119 Nissan and its ally Renault: Hans Greimel, "Renault, Nissan to Go Electric, Seek Leadership in Green Cars," *Automotive News Europe*, February 4, 2008.

121 "Thousands of persons have large pecuniary interests": Quoted in Jonnes, *Empires of Light*, p. 202.

122 Although Toyota had announced: Alan Ohnsman and Jeff Plungis, "Toyota Questions Cost, Batteries of Plug-In Hybrids," *Bloomberg*, May 18, 2009.

122 a car for "idiots": Lawrence Ulrich, "Audi President Has Verbal Jolt for Volt," MSN Autos, September 2, 2009, http://editorial.autos.msn.com/blogs/autosblogpost .aspx?post=1247701.

124 ten things Letterman should know: Mark Phelan, "Hey, Letterman: Here Are 10 Things About the Volt," *Detroit Free Press*, May 20, 2009.

126 Fritz Henderson was out: Bill Vlasic, "In the Changeover at G.M., a New Hands-On Attitude," *New York Times*, December 14, 2009.

128 They had allies in Warren Buffett: Keith Bradsher, "Buffett Buys Stake in Chinese Battery Manufacturer," *New York Times*, September 29, 2008.

128 world's largest supplier of electric cars by 2012: Keith Bradsher, "China Vies to Be World's Leader in Electric Car," *New York Times*, April 1, 2009.

8: The Stimulus

131 William Clay Ford, Jr.: Bill Vlasic, "Ford Scion Looks Beyond Bailout to Green Agenda," *New York Times*, November 24, 2008.

132 Rick Wagoner, still at the helm of GM: Sholnn Freeman, "GM Says Batteries for Volt Might Not Be U.S. Produced," *Washington Post*, September 18, 2008.

132 Charles Gassenheimer spoke: David E. Zoia, "All Charged Up," *Ward's Auto World*, November 1, 2008.

133 "This is the single largest investment": Official transcript of Steven Chu's remarks, www.energy.gov/7751.htm.

139 "hottest IPO of 2009": "Cramer's IPO Play—A123 Systems," http://maddmoney .net/cramers-ipo-play-a123-systems/.

139 the day A123 went public: Erin Ailworth, "IPO Fuels Prospects of Battery Makers," *Boston Globe*, September 25, 2009.

142 "We ended up having to teach these guys": Don Lee, "Battery Recharges Debate About U.S. Manufacturing," *Chicago Tribune*, May 16, 2010.

143 In May 2008, Sanyo: "Volkswagen, Sanyo to Develop Lithium-Ion Batteries," *Reuters*, May 28, 2008, www.reuters.com/article/idUST26662320080528.

144 That July, Panasonic announced: "Panasonic to Build Lithium-ion Battery Plant in Osaka," *JCN Newswire*, www.japancorp.net/article.asp?Art_ID=19219.

144 Mitsubishi, through a joint venture: "Lithium Energy Japan Secures Plant Site and Buildings for World's First Mass Production of Large Lithium-ion Batteries for EVS," Lithium Energy Japan press release, www.gsyuasa-lp.com/News/LEJ _20080806e.pdf.

144 Nissan's joint venture with NEC: "Nissan and NEC Joint Venture—AESC— Starts Operations," Nissan press release, www.nissan-global.com/EN/NEWS/ 2008/_STORY/080519-01-e.html.

144 Charles Gassenheimer cited Nissan's recent purchase: "Ener1, Inc. at Jefferies & Co. Global Clean Technology Conference," FD (Fair Disclosure) Wire, October 22, 2008.

144 LG Chem planning to devote a cell factory: John Voelcker, "Global Market Review of Hybrids and Electric-Drive Vehicles—Forecasts to 2015," *Just-Auto*, April 2009.

144 Samsung was also launching an automotive joint venture: David E. Zoia, "All Charged Up," *Ward's Auto World*, November 1, 2008.

144 BYD's thirty thousand workers: Amy Hsuan, "Governor Sees Hybrid as Green Fit for Oregon," *Oregonian*, November 23, 2008.

145 "This appears to be the kind of deal": Taro Fuse and Kentaro Hamada, "Panasonic in Tentative Deal to Buy Sanyo," *New York Times*, November 2, 2008.

9: The Prospectors

150 In December 2006: William Tahil, "The Trouble with Lithium: Implications of Future PHEV Production for Lithium Demand," http://tyler.blogware.com/ lithium_shortage.pdf.

150 Earlier that year: "Ground Zero: The Nuclear Demolition of the World Trade Centre," www.nucleardemolition.com/GZero_Sample.pdf.

151 Tahil was quoted in mainstream publications: Brendan I. Koerner, "The Saudi

Arabia of Lithium," *Forbes*, November 2008; Rebecca Coons, "Lithium: Charging Up the Hybrids," *Chemical Week*, July 22, 2008.

152 Evans wrote his own estimate: R. Keith Evans, "An Abundance of Lithium," March 2008, www.che.ncsu.edu/ILEET/phevs/lithium-availability/An_Abundance _of_Lithium.pdf.

153 Tahil hit back: William Tahil, "The Trouble with Lithium 2: Under the Microscope," www.meridian-int-res.com/Projects/Lithium_Microscope.pdf.

153 He posted a response paper online: R. Keith Evans, "An Abundance of Lithium Part Two," July 2008, www.evworld.com/library/KEvans_LithiumAbunance_pt2.pdf.

155 a company called Nordic Mining: All remaining information in this paragraph comes from Keith Evans's presentation the first day of the Lithium Supply and Markets conference, 2010.

157 some 8.9 million tons: R. Keith Evans, "Lithium Reserves and Resources," paper presented at Lithium Supply and Markets 2010, Las Vegas, January 2010.

162 After the volcano's collapse: See John McPhee's *Annals of the Former World* for an accessible explanation of basin-and-range faulting in this corner of Nevada.

10: The Lithium Triangle

168 starting in 2009, the Salar de Uyuni: Simon Romero, "In Bolivia, Untapped Bounty Meets Nationalism," *New York Times*, February 3, 2009; Eitan Haddock, "En Bolivie, la Ruée Vers L'or Gris," *Le Monde Magazine*, December 5, 2009; Jeffrey Kofman, "Bolivia's Uyuni Salt Flats Hold Promise of Greener Future," ABC News, August 5, 2009, http://abcnews.go.com/Technology/JustOneThing/story?id =8257028&page=1; Peter Day, "Battery Power," BBC, January 9, 2009, www.bbc .co.uk/worldservice/business/2009/09/090901_globalbusiness_010909.shtml.

168 Bolivia's gross domestic product: *The CIA World Factbook*: Bolivia. www.cia.gov/ library/publications/the-world-factbook/geos/bl.html.

168 ExxonMobil's profit: Jad Mouawad, "Exxon Grew as Oil Industry Contracted," *New York Times*, February 1, 2010.

168 Since gaining independence: See Chasteen, *Born in Blood and Fire*, for a history of Bolivia.

169 The most notorious example: Lawrence Wright, "Lithium Dreams," *New Yorker*, March 22, 2010.

169 self-described yuppie: William Finnegan, "Leasing the Rain," *New Yorker*, April 8, 2002.

170 his "campaign manager": Jeremy D. Rosner and Mark Feierstein, "Hindering Reform in Latin America," *Washington Post*, August 6, 2002.

170 the military killed six Aymara villagers: Grace Livingstone, "International Roundup: Americas: Unions in Bolivia Go on Strike," *Guardian*, September 29, 2003.

170 The military responded violently in El Alto: Anthony Faiola, "Ex-President of Bolivia Faces Suit in U.S.," *Washington Post*, September 26, 2007.

170 Carlos Mesa, broke with the government: Larry Rohter, "Bolivian President Remains Defiant as Protests Intensify," *New York Times*, October 14, 2003.

170 Goni resigned and fled the country: Larry Rohter, "Bolivian Leader Resigns and His Vice President Steps In," *New York Times*, October 18, 2003.

170 in 2005 the government passed a law: Juan Forero, "Foreign Gas Companies in Bolivia Face Sharply Higher Taxes," *New York Times*, May 18, 2005.

171 five hundred years of resistance: Jeremy McDermott, "Bolivian Leader Sworn in as the Left Advances on US Doorstep," *Daily Telegraph* (London), January 23, 2006.

171 "Death to the Yankees": See the documentary *Cocalero* (2007), directed by Alejandro Landes.

171 stopped cooperating with the U.S. Drug Enforcement Agency: "Bolivia Suspends U.S.-Backed Antidrug Efforts," *New York Times*, November 1, 2008.

171 "the ambassador of the United States is conspiring": Jeremy McDermott, "Bolivia Expels US Ambassador Philip Goldberg," *Telegraph*, September 12, 2008.

171 Mitsubishi, Sumitomo, the Chinese government: Candace Piette, "Bolivians Learn Chinese to Boost Their Trade Options," BBC News, December 24, 2009.

171 Evo Morales traveled to Paris: Jenny Barchfield, "Bolivia's Morales Calls on Total to Up Investment," Associated Press, February 17, 2009.

172 "Andean capitalism": "Bolivia: A New Phase Begins," *Socialism Today*, February 2006.

172 extracts boron from a salar: Tierra website, www.tierra.bo/es/tierra.php.

172 Tierra found itself in the middle of a firestorm: Dominican Network: The Delegation of the Order of Preachers to the United Nations, August 10, 2003, "Item 3: Administration of Justice—Bolivia."

172 Roelants got twelve years: "Belgium 'Astonished' at 12-Year Sentence for Belgian Citizen in Bolivia," BBC Summary of World Broadcasts, July 22, 2003, www.accessmylibrary.com/coms2/summary_0286-23904531_ITM.

173 Belgian "Rasputin": Juan Carlos Zuleta, "Retrasos y Posible Conflicto de Intereses Empañan el Proyecto Piloto de Litio," *El Potosí*, April 23, 2010, www.elpotosi.net/noticias/2010/0423/noticias.php?nota=23_04_10_opin3.php.

173 That morning in *La Razón*: Ramiro Prudencio Lizón, "Crisis de la Minería en Bolivia," *La Razón*, April 28, 2010.

174 "Bolivian President Evo Morales' decision": Harvey Beltrán, "Analysis: Lithium Development—One Step Forward, Two Steps Back—Bolivia," *Business News Americas*, March 31, 2010.

175 villagers who live near the San Cristóbal mines: "As Protests Mount Against San Cristóbal Silver Mine, Bolivia Looks to Extract Massive Lithium Reserves, But at What Cost?" Democracy Now, April 20, 2010, www.democracynow.org/2010/4/20/two.

175 that Guillermo Roelants be expelled: "Intereses Políticos y Económicos Anidan en el Sudoeste," *El Potosí*, April 19, 2010.

175 120,000 tons of lithium carbonate equivalent: Roskill Information Services, "The

Lithium Market: 2009 Review and Outlook," paper presented at the Lithium Supply and Markets conference, Las Vegas, January 26–28, 2010.

175 Lithium-demand forecasts get hazy: Ibid.

176 some nine million tons of lithium: USGS 2010 Lithium Report, http://minerals .usgs.gov/minerals/pubs/commodity/lithium/mcs-2010-lithi.pdf.

178 "Bolivia has the largest resources": Quoted on the Comibol website, www .evaporiticosbolivia.org/indexi.php.

180 Because of the constant influx of minerals: François Risacher and Bertrand Fritz, "Quaternary Geochemical Evolution of the Salars of Uyuni and Coipasa, Central Altiplano, Bolivia," *Chemical Geology* 90 (1991): 211–31.

182 the driest place on earth: Priit J. Vesilind, "The Driest Place on Earth," *National Geographic*, August 2003.

182 bacteria-detecting Mars robots: Michael Coren, "Digging for Life in the Deadest Desert: Driest Spot on Earth May Hold Clues to Mars," CNN, August 5, 2004, www.cnn.com/2004/TECH/space/08/04/atacama.desert/index.html.

182 since before the arrival of the Spanish: Kevin J. Vaughn, Moises Linares Grados, Jelmer W. Eerkens, and Matthew J. Edwards, "Hematite Mining in the Ancient Americas: Mina Primavera, A 2,000 Year Old Peruvian Mine," *Journal of the Minerals, Metals and Materials Society* 59 (2007): 16–20.

182 just about supported the entire country: Chasteen, *Born in Blood and Fire*, p. 178.

184 The War of the Pacific: Ibid., pp. 178–79.

186 the beneficiary of blatant nepotism: The accusations are well documented. Sources used in the research for this book include Timothy L. O'Brien and Larry Rohter, "The Pinochet Money Trail," *New York Times*, December 12, 2004; Gabriel Agosin O., "Un Intocable en el Banquillo," *La Nación*, August 7, 2005 (which contains the Kaizer Soze reference mentioned in the text). Ponce Lerou receives two mentions in Pamela Constable and Arturo Valenzuela's fascinating history of Chile under the rule of Pinochet, *A Nation of Enemies*, pp. 74, 216.

189 Roelants gave an interview: "Delays and Possible Conflict of Interests Cloud Bolivia's Lithium Pilot Project," Juan Carlos Zuleta's Instablog, April 24, 2010.

190 Pieces published in *El Potosí* and *El Diario*: Zuleta, "Retrasos y Posible Conflicto de Intereses"; "Oficinas de EMBRE Deben Tener Residencia en Lípez," *El Diario*, April 30, 2010.

190 I had seen his picture in *La Razón*: Víctor Quintanilla, "Bolloré Pide un Acuerdo Sobre el Litio 'Lo Más Pronto Posible,'" *La Razón*, April 28, 2010.

II: The Goal

196–97 Panasonic will start selling in 2013: Hideyoshi Kume, "Panasonic's New Li-Ion Batteries Use Si Anode for 30% Higher Capacity," *Nikkei Electronics Asia*, March 1, 2010.

199 In 2009, Sadoway received a $7 million grant: "Bold, Transformational Energy Research Projects Win $151 Million in Funding," Advanced Research Projects

Agency, U.S. Department of Energy news release, October 26, 2009, http://arpa-e .energy.gov/Media/News/tabid/83/vw/1/ItemID/20/Default.aspx.

200 Lithium-air has a theoretical maximum: S. J. Visco, E. Nimon, and L. C. De Jonghe, "Lithium-Air," *Encyclopedia of Electrochemical Power Sources* (New York: Elsevier, 2009).

SELECTED BIBLIOGRAPHY

Boschert, Sherry. *Plug-in Hybrids: The Cars That Will Recharge America*. Gabriola Island, BC: New Society Publishers, 2006.

Cade, John F. J. "Lithium Salts in the Treatment of Psychotic Excitement." *Medical Journal of Australia* 2, no. 36 (1949): 349–52.

Chasteen, John Charles. *Born in Blood and Fire: A Concise History of Latin America*. 2nd ed. New York: Norton, 2006.

Chung, Sung-Yoon, and Yet-Ming Chiang, "Microscale Measurements of the Electrical Conductivity of Doped $LiFePO_4$." *Electrochemical and Solid-State Letters* 6(12) (2003): A278–81.

Chung, Sung-Yoon, Jason T. Bloking, and Yet-Ming Chiang. "Electronically Conductive Phospho-olivines as Lithium Storage Electrodes." *Nature Materials* 1 (2002): 123–28.

———. "From Our Readers: On the Electronic Conductivity of Phospho-olivines as Lithium Storage Electrodes." *Nature Materials* 2 (2003): 702–703.

Constable, Pamela, and Arturo Valenzuela. *A Nation of Enemies: Chile Under Pinochet*. New York: Norton, 1991.

Deffeyes, Kenneth S. *Hubbert's Peak: The Impending World Oil Shortage*. Princeton: Princeton University Press, 2001.

Delacourt, Charles, Philippe Poizot, Jean-Marie Tarascon, and Christian Masquelier. "The Existence of Temperature-Driven Solid Solution in Li_xFePo_4 for $0 \leq x \leq 1$." *Nature Materials* 4 (2005): 254–60.

Doyle, Jack. *Taken for a Ride: Detroit's Big Three and the Politics of Pollution*. New York: Four Walls Eight Windows, 2000.

Ellis, Brian, P. Subramanya Herle, Y.-H. Rho, Linda F. Nazar, R. Dunlap, Laura K. Perry, and D. H. Ryan. "Nanostructured Materials for Lithium-ion Batteries: Surface Conductivity vs. Bulk Ion/Electron Transport." *Faraday Discussions* 134 (2007): 119–41.

Ellis, Brian, Marnix Wagemaker, Fokko M. Mulder, and Linda F. Nazar. "Comment on 'Aliovalent Substitutions in Olivine Lithium Iron Phosphate and Impact on Structure and Properties.'" *Advanced Functional Materials* 20 (2010): 186–88.

El-Mallakh, Rif S., and James W. Jefferson. "Prethymoleptic Use of Lithium." *American Journal of Psychiatry* 156, no. 1 (1999): 129.

El-Mallakh, Rif S., and Rona Jeannie Roberts. "Lithiated Lemon-Lime Sodas." *American Journal of Psychiatry* 164, no. 11 (2007): 1662.

Goodenough, John B. *Witness to Grace.* Baltimore: PublishAmerica, 2008.

Halberstam, David. *The Reckoning.* New York: Morrow, 1986.

Holstein, William J. *Why GM Matters: Inside the Race to Transform an American Icon.* New York: Walker, 2009.

Huggins, Robert Alan. *Advanced Batteries: Materials Science Aspects.* New York: Springer, 2009.

Ingrassia, Paul. *Crash Course: The American Automobile Industry's Road from Glory to Disaster.* New York: Random House, 2010.

Jacobs, Chip, and William J. Kelly. *Smogtown: The Lung-Burning History of Pollution in Los Angeles.* Woodstock, NY: Overlook Press, 2008.

Jonnes, Jill. *Empires of Light: Edison, Tesla, Westinghouse, and the Race to Electrify the World.* New York: Random House, 2004.

Josephson, Matthew. *Edison: A Biography.* New York: Wiley, 1992.

Kirsch, David A. *The Electric Vehicle and the Burden of History.* New Brunswick, NJ: Rutgers University Press, 2000.

Makimura, Y., L. S. Cahill, Y. Iriyama, G. R. Goward, and L. F. Nazar. "Layered Lithium Vanadium Fluorophosphate, $Li_5V(PO_4)_2F_2$: A 4 V Class Positive Electrode Material for Lithium-Ion Batteries." *Chemistry and Materials* 20 (2008): 4240–48.

McPhee, John. *Annals of the Former World.* New York: Farrar, Straus and Giroux, 1998.

Meethong, Nonglak, Yu-Hua Kao, Scott A. Speakman, and Yet-Ming Chiang. "Aliovalent Substitutions in Olivine Lithium Iron Phosphate and Impact on Structure and Properties." *Advanced Functional Materials* 19 (2009): 1060–70.

Mizushima, K., P. C. Jones, P. J. Wiseman, and J. B. Goodenough, "Li_xCoO_2 ($0<x<-1$): A New Cathode Material for Batteries of High Energy Density." *Materials Research Bulletin* 15, no. 6 (1980): 783–89.

Moore, Gregory J., Joseph M. Bebchuk, Ian B. Wilds, Guang Chen, and Husseini K. Menji. "Lithium-Induced Increase in Human Brain Grey Matter." *Lancet* 356 (2000): 1241–42.

Nelson, Paul A., Danilo J. Santini, and James Barnes. "Factors Determining the Manufacturing Costs of Lithium-Ion Batteries for PHEVs." *EVS24 International Battery, Hybrid and Fuel Cell Electric Vehicle Symposium* (2009): 1–12.

Ohgami, Hirochika, Takeshi Terao, Ippei Shiotsuki, and Nobuyoshi Ishii. "Lithium Levels in Drinking Water and Risk of Suicide." *British Journal of Psychiatry* 194 (2009): 464–65.

Padhi, A. K., K. S. Nanjundaswamy, and J. B. Goodenough. "LiFePO$_4$: A Novel Cathode Material for Rechargeable Batteries." *Electrochemical Society Meeting Abstracts* 96 (1996): 73.

———. "Phospho-olivines as Positive-Electrode Materials for Rechargeable Lithium Batteries." *Journal of the Electrochemical Society* 144 (1997): 1188–94.

Pancaldi, Giuliano. *Volta: Science and Culture in the Age of Enlightenment.* Princeton: Princeton University Press, 2003.

Panel on Electrically Powered Vehicles, U.S. Department of Commerce. *The Automobile and Air Pollution: A Program for Progress.* Washington, D.C.: GPO, 1967.

Pooley, Eric. *The Climate War: True Believers, Power Brokers, and the Fight to Save the Earth.* New York: Hyperion, 2010.

Powers, William. *Whispering in the Giant's Ear: A Frontline Chronicle from Bolivia's War on Globalization.* New York: Bloomsbury, 2006.

Ravet, Nathalie, Ali Abouimrane, and Michel Armand. "From Our Readers: On the Electronic Conductivity of Phospho-olivines as Lithium Storage Electrodes." *Nature Materials* 2 (2003): 702–703.

Ravet, Nathalie, Y. Chouinard, J. F. Magnan, S. Besner, M. Gauthier, and M. Armand. "Electroactivity of Natural and Synthetic Triphylite." *Journal of Power Sources* 97–98 (2001): 503–507.

Ravet, Nathalie, J. B. Goodenough, S. Besner, M. Simoneau, P. Hovington, and M. Armand. "Improved Iron Based Cathode Material." Paper presented at the 196th meeting of the Electrochemical Society, Honolulu, October 17–22, 1999.

Ravet, Nathalie, J-F. Magnan, J. M. Gauthier, and M. Armand. "Lithium Iron Phosphate: Towards an Universal Electrode Material." Paper presented at the International Conference on Materials for Advanced Technologies, Singapore, July 1–6, 2001.

Sampson, Anthony. *The Seven Sisters: The Great Oil Companies and the World They Shaped.* New York: Viking, 1975.

Schallenberg, Richard H. *Bottled Energy: Electrical Engineering and the Evolution of Chemical Energy Storage.* Philadelphia: American Philosophical Society, 1982.

Schiffer, Michael B., Tamara C. Butts, and Kimberly K. Grimm. *Taking Charge: The Electric Automobile in America.* Washington, D.C.: Smithsonian Institution Press, 1994.

Schlesinger, Henry. *The Battery: How Portable Power Sparked a Technological Revolution.* Washington, D.C.: Smithsonian Books, 2010.

Shacket, Sheldon R. *The Complete Book of Electric Vehicles.* 2nd ed. Northbrook, IL.: Domus Books, 1981.

Shnayerson, Michael. *The Car That Could: The Inside Story of GM's Revolutionary Electric Vehicle.* New York: Random House, 1996.

Subramanya Herle, P., B. Ellis, N. Coombs, and L. F. Nazar. "Nano-network Electronic Conduction in Iron and Nickel Olivine Phosphates." *Nature Materials* 3 (2004): 147–52.

Thackeray, Michael. "Lithium-ion Batteries: An Unexpected Conductor." *Nature Materials* 1 (2002): 81–82.

Thackeray, Michael, W.I.F. David, P. G. Bruce, and J. B. Goodenough. "Lithium Insertion into Manganese Spinels." *Materials Research Bulletin* 18, no. 4 (1983): 461–72.

van Gool, W., ed. *Fast Ion Transport in Solids: Solid State Batteries and Devices.* New York: Elsevier, 1973.

Whittingham, M. S. "Electrical Energy Storage and Intercalation Chemistry." *Science* 192, no. 4244 (1976): 1126–27.

Whittingham, M. S., and R. A. Huggins. "Beta Alumina: Prelude to a Revolution in Solid State Electrochemistry." In *Solid State Chemistry: Proceedings of the 5th Materials Research Symposium,* edited by Robert S. Roth and Samuel J. Schneider Jr., 139–54. Washington, D.C.: GPO, 1972.

Xu, Yong-Nian, Sung-Yoon Chung, Jason T. Bloking, Yet-Ming Chiang, and W. Y. Ching. "Electronic Structure and Electrical Conductivity of Undoped LiFePO$_4$." *Electrochemical and Solid-State Letters* 7 (2004): A131–34.

Yergin, Daniel. *The Prize: The Epic Quest for Oil, Money, and Power.* New York: Free Press, 2008.

Young, Allan H. "Invited Commentary on . . . Lithium Levels in Drinking Water and Risk of Suicide." *British Journal of Psychiatry* 194 (2009): 466.

ACKNOWLEDGMENTS

Somewhere in the reporting of this book I lost count of the scientists, engineers, industrialists, entrepreneurs, analysts, and other knowledgeable guides who took the time to speak to me. Far more people than are named or quoted in these pages had an influence. Instead of attempting and inevitably failing to type up a complete list, I'd like to issue a blanket declaration of gratitude to every person who met with me or got on the phone to explain his or her area of expertise. I hope I did your subjects justice.

I doubt I would ever have set out to write this book if it hadn't been for early encouragement from friends. I'm grateful to Gabe Sherman for one evening saying: Have you ever thought about writing a book about this stuff? Thanks to Claire Martin for urging me to go for it, and to Christian Debenedetti for moral support throughout the process. Arianne Cohen was an essential early adviser.

Larry Weissman, my agent, has been an indispensable ally. He saw clearer promise in my germ of an idea than I did, and after coaching me through the writing of the proposal, he got it on the desk of the person who finally made it happen: Joe Wisnovsky, my wise and unfailingly supportive editor at Hill and Wang. He and the rest of the professionals at Hill and Wang / Farrar, Straus and Giroux have been a pleasure to work with.

At *Popular Science*, Mark Jannot, Jake Ward, and Mike Haney were all extraordinarily accommodating as I worked on this book. During my travels I relied on favors from more people than I can count, but in South America the assistance of locals was particularly crucial. In Bolivia, Francisco Quisbert and our guide, Enoch, were tremendously helpful. In Chile, Andres Yaksic went some eight hundred miles out of his way to get me behind the gates of SQM.

Abe Streep, Chelsea Sexton, Eddie Alterman, and Jeff Dahn each supplied critical feedback on portions of this book. Many thanks to Michele Gardner for her help with legal research. I'm grateful to Chris Canipe for providing the map of the Lithium Triangle, and to Jennifer Stahl for the amazingly sharp and careful fact-checking she performed on certain thorny passages. I remain responsible for any and all errors that made it into print.

For endless support over the years I want to thank my family: Mom, Dad, Luke, grandmas, grandpas, aunts, uncles, and these days, my in-laws. And it's hard to know where to begin thanking Leigh Garrison-Fletcher, my wife and best friend. I spent many evenings and weekends either working on or brooding about this book, and she tolerated them all. On the South American leg of my reporting, she was, as always, my ideal travel companion. Back at home, she read and edited drafts, helped assemble endnotes, and generally took care of maintaining our lives, all while making my life wonderful.

INDEX

Page numbers in *italics* refer to maps.